The Twilight of Evolution

The Twilight of Evolution

Henry M. Morris

BAKER BOOK HOUSE
Grand Rapids, Michigan

Library of Congress Catalog Card Number: 63-21471

ISBN: 0-8010-5862-7

First printing, November 1963
Second printing, May 1964
Third printing, October 1964
Fourth printing, April 1965
Fifth printing, January 1966
Sixth printing, November 1966
Seventh printing, October 1967
Eighth printing, May 1968
Ninth printing, May 1969
Tenth printing, September 1970
Eleventh printing, August 1971
Twelfth printing, February 1972
Thirteenth printing, May 1972
Fourteenth printing, June 1973
Fifteenth printing, May 1974
Sixteenth printing, May 1975
Seventeenth printing, April 1976
Eighteenth printing, April 1977
Nineteenth printing, September 1978
Twentieth printing, September 1980

PHOTOLITHOPRINTED BY CUSHING - MALLOY, INC.
ANN ARBOR, MICHIGAN, UNITED STATES OF AMERICA

Preface

Probably the most important single issue confronting Biblical Christianity in these days is the question of origins. The remaining strongholds of virile, Bible-centered Christian witness in the world seem everywhere to be in imminent and serious danger of capitulation to the forces of philosophical evolution. The evolutionary perspective has captured the intellectual world and is rapidly overwhelming the religious world as well. The Biblical revelation of special creation is rarely accepted any longer as an actual historical account of the events of God's creation. Even in those religious circles still regarded as theologically conservative, the Bible record is increasingly being interpreted as "allegorical" or "liturgical," rather than as historical and scientific.

Many people have no concern over this struggle, regarding the issue as "peripheral." The important thing, it is alleged, is simply to acknowledge the fact of God and of "creation," leaving the problem of the method and sequence of creation to be worked out by the scientists.

Although the legendary posture of the ostrich may provide a superficial and temporary freedom from the discomforts of active conflict, it is not to be recommended as a permanent solution! The leaders of evolutionary thought in the intellectual world, and most of their followers, are not to be satisfied with any such shallow compromise as that which regards evolution as "God's method of creation." If evolution can explain the origin and development of this universe and its inhabitants, then there is no need for any kind of personal God at all.

The tale could be told a thousand times, of a Christian church or school or mission society or some other organization, founded by men of strong Biblical faith and with an uncompromising evangelical witness, slowly but steadily drifting off its foundations and gradually sinking in the sands of modernism and secularism. This tragedy, repeated times without number, almost always begins with a questioning of Biblical creationism. The Scriptural account of origins must somehow be accommodated to the latest scientific theories of origins (which are always evolutionary). This accommodation inevitably and necessarily leads to a softening of the doctrine of Biblical inspiration and infallibility. Other creative acts of God (that is, the recorded miracles) begin to be questioned, and a view of Biblical inspiration which allows for cultural limitations and even for outright contradictions becomes adopted. The authority of the Gospels and even of Jesus Christ then must necessarily be questioned. Soon the entire Bible becomes merely a record of man's religious evolution, an outgrowth of his biological evolution. The proper activity for modern Christians therefore eventually becomes mere "social action," striving to help in the future evolution of the social order into a more advanced and enlightened humanistic society.

For what a man believes about ultimate origins and about God's revelation concerning creation will inevitably affect his beliefs concerning destinies and purposes. The issue of evolution versus Biblical creation is most emphatically *not* a peripheral question. In view of widespread defections, in very recent years, of once-sound schools and other organizations to the evolutionary forces, it has become rather an issue of the gravest urgency.

This little book represents an attempt to delineate this issue in terms of its ultimate implications and outcome. It is an expansion of two lectures, with their attendant discussions, given in November 1962, at Grand Rapids, Michigan, under sponsorship of the Reformed Fellowship, Inc., a group of Christian believers who hold to the Reformed Faith and express their adherence to the Calvinistic creeds as formulated in the Belgic Confession, the Heidelberg Catechism, the Canons of Dort, and the Westminster Confession and Catechism. Their purpose is to give sharpened expression to this faith, to stimulate doctrinal sensitivities of those who prefer this faith, and to promote the spiritual welfare of the Reformed churches everywhere. The author desires to express his deepest appreciation to these men, not only for their interest in sponsoring these lectures and this book, but even more for the warmth of their true Christian fellowship and for their uncompromising stand for Bible Christianity in these critical days.

Sincere appreciation is also expressed to two colleagues, Dr.

John C. Whitcomb, Jr., and Dr. Walter E. Lammerts. These men have reviewed the manuscript and their very helpful suggestions have been incorporated in the book. Dr. Whitcomb is Professor of Old Testament at Grace Theological Seminary in Winona Lake, Indiana, and is co-author of *The Genesis Flood,* to which this present book serves in part as a sequel. Dr. Lammerts is a biologist, specializing in horticultural genetics, and is also an amateur geologist. He is Chairman of the recently formed Creation Research Committee, of which the author is privileged to be a member, along with a number of other Christian scientists who are specialists in the various disciplines concerned with this question of origins. Of course, although the help and encouragement of the Reformed Fellowship and of Doctors Whitcomb and Lammerts is most gratefully acknowledged, the author emphasizes that he alone assumes responsibility for the book and its emphasis. Finally, a special word of thanks is due the author's mother, Mrs. I. H. Morris, who graciously typed both the preliminary and final drafts of the manuscript.

It is the prayer of all of us that God may use this little book "as seemeth him good." We would, in so far as our limited comprehension of that which is spiritual and eternal may permit, unite in spirit with the elders around the throne of him that liveth forever and ever and say with them:

> "Thou art worthy, O Lord, to receive glory and honor and power: for thou hast created all things, and for thy pleasure they are and were created" (Revelation 4:11).

Henry M. Morris

Blacksburg, Virginia
March, 1963

Foreword

The Reformed Fellowship, Inc. finds a large measure of satisfaction in publishing this book. This satisfaction stems from a number of considerations.

The author's steadfast allegiance to the Holy Scriptures is reason for highest satisfaction. He would be strictly loyal to God's Word of Truth. It is refreshing to see such loyalty to God's Word on a high level. Naturally our admiration for this element in the book does not call for full agreement with the author at every point of exegesis.

A second source of satisfaction is found in the scholarship this little book reveals. The author has obviously read widely in the literature on the subject, and he buttresses his argument constantly with effective citations from a rather amazing variety of reputable sources. It is a pleasure to present a book with this commendable feature.

Furthermore, deep gratification is afforded by the fact that the author, himself a reputable scientist in his field, has a sharp understanding of the important philosophical and spiritual issues involved in evolutionary thinking and writing. He does a particularly effective job in unmasking the central philosophical and spiritual issue that lurks behind the often impressive front of scientific respectability of so much evolutionary writing and propaganda.

It is a pleasure, therefore, to present an up-to-date, Bible-oriented, and scientifically pertinent discussion on a strategically important theme within relatively brief compass. This book should be widely read for these reasons. May it enjoy a wide circulation and may it give to the discussion of this important subject a tone and direction that will call the minds and hearts of many back to what God has written.

Reformed Fellowship, Inc.

Contents

Chapter I

The Influence of Evolution

In a new book dealing with evolution, John T. Bonner, one of the nation's leading biologists, says:

> "....all biologists, I think, would agree that evolution is the largest and most encompassing [theme] of them all. Evolution has provided the framework for life in general, and therefore it will be the theme of this book."[1]

With respect to the Darwinian doctrine of evolution by chance variation and natural selection, he says:

> "When Darwin published his book, it created a tremendous sensation; it seemed to be the key for which the whole world was waiting.... In the hundred years since the publication of *Origin of Species,* our opinion of Darwin was never so high as it is now."[2]

In view of such glowing present-day estimates of the concept of Darwinian evolution, opinions shared without doubt by a substantial majority of modern American biologists, to entitle a book "The Twilight of Evolution" may sound like wishful thinking at best. As a matter of fact, it is not intended to suggest at all that the evolutionary philosophy is about to be discarded in favor of creation by any great number of biologists; this is decidedly not the case.

But it does seem to be increasingly clear that evolution is not a *science.* Evidence continues to accumulate that it is rather an anti-Christian, anti-theistic way of thought, a system rather than

---

[1] *The Ideas of Biology* (New York; Harper & Brothers, 1962), p. ix.

[2] *Ibid,* p. 45. Dr. Bonner is Professor of Biology at Princeton.

a science, a philosophy instead of a history. The issues are becoming more sharp and the lines more distinctly drawn, so that there is less reason now than even a decade ago for the Biblical Christian to seek hermeneutical compromises with evolution. The known facts of science and history can be interpreted in either of two frameworks – that of evolutionary progress or that of Creation and the Fall, and the framework chosen depends not on inductions from the known facts, but rather on deductions from one's basic presuppositions.

It is contended in this little book, therefore, that evolution is rapidly declining in status with men whose presuppositions are Bible-centered. It is no longer a fearful foe from which we must retreat or with which we must compromise. Its nature is coming more plainly into focus and can be discerned as that of a vast framework of deductions built upon the foundation of a false premise. The false premise is the blasphemous assumption that man, who is a creature of God – a fallen creature at that – can explain God's creation without God and His revealed Word. Thus he worships and serves the creature rather than the Creator. Like his father Adam, when he has rejected God's Word, he seeks to hide from God, and thus must explain everything in the universe, if he can, without reference to its Creator.

The most remarkable paradox of modern scientism is that the system of evolution could ever have obtained such nearly universal acceptance while being so utterly devoid of any genuine scientific basis. The actual nature of the pertinent scientific data, and the reasons for this widespread delusion, will be explained in later chapters. In this chapter, we shall examine in some detail the extent of the influence of evolution in modern thought.

We have already noted that total evolution is almost universally accepted by biologists. Sir Julian Huxley, the famous British biologist (and, incidentally, the first Director-General of UNESCO), has emphasized the all-inclusive nature of the evolutionary philosophy:

> "The concept of evolution was soon extended into other than biological fields. Inorganic subjects such as the life-histories of stars and the formation of the chemical elements on the one hand, and on the other hand subjects like linguistics, social

anthropology, and comparative law and religion, began to be studied from an evolutionary angle, until today we are enabled to see evolution as a universal and all-pervading process."[3]

A little later, in almost a spirit of missionary fervor, Huxley exclaims:

"Furthermore, with the adoption of the evolutionary approach in non-biological fields, from cosmology to human affairs, we are beginning to realize that biological evolution is only one aspect of evolution in general. Evolution in the extended sense can be defined as a directional and essentially irreversible process occurring in time, which in its course gives rise to an increase of variety and an increasingly high level of organization in its products. Our present knowledge indeed forces us to the view that the whole of reality *is* evolution—a single process of self-transformation."[4]

Huxley has exerted a profound influence on 20th-century thought even as his grandfather, Thomas Huxley, did in the 19th century. And if evolution is what he claims it is, it essentially has taken the place of God himself!

That his estimate of the importance of the evolutionary philosophy has not been exaggerated can be verified by examining the writings of the leaders in all the various natural and social sciences. In astronomy and cosmology, for example, it is all but universally held that everything in the physical universe has gradually, over billions of years, evolved mechanically into its present state. Although cosmological theories are numerous, and new ones appear frequently, all are agreed that the universe is evolving:

"We should expect to find stars and galaxies in all stages of evolution as they form from existing material and then decay. For stars this is certainly the case. . . . There may be a similar evolutionary process for galaxies, but at the moment we do not have enough experimental evidence to give us the clues to the evolutionary pattern."[5]

---

[3] "Evolution and Genetics," Chapter 8 in *What is Science?* (Ed., J. R. Newman, New York; Simon and Schuster, 1955) p. 272.

[4] *Ibid,* p. 278.

[5] Gerald S. Hawkins: "A New Theory of the Universe," *Science Digest,* (Vol. 52, November, 1962), p. 45. Hawkins is Professor of Astronomy at Boston University.

The last statement is an encouragingly humble admission. After all, one would anticipate some difficulty in obtaining *experimental* evidence on the evolution of galaxies!

With respect to the origin of the earth and the solar system, there similarly are numerous theories extant, all of which (except the true account, in the Bible), are evolutionary in nature. As Isaac Asimov says:

> "In the 19th century, as concepts of long-drawn-out natural processes such as Hutton's uniformitarian principle won favor, catastrophes went out of fashion. Instead, scientists turned more and more to theories involving evolutionary processes, following Newton rather than the Bible."[6]

In like fashion, the study of the earth and its history is oriented entirely in terms of an evolutionary perspective. In fact, it is well known that the very basis for the subdivision of periods of geologic history is evolutionary in nature. That is, rock formations are identified in terms of the evolutionary succession of life as it developed during the time of geologic history when the rocks were deposited. As the Yale geologist, Carl O. Dunbar, says:

> "We now know, of course, that different kinds of animals and plants have succeeded one another in time because *life has continually evolved;* and inasmuch as organic evolution is world-wide in its operation, only rocks formed during the same age could bear identical faunas."[7]

This particular aspect of the problem of evolution is highly important, as we shall see later. However, if evolution were merely a scientific theory affecting the interpretation of the data of biology, geology, and astronomy, we would not be too concerned about it. Assuming that the problem of harmonizing evolutionary history with the Biblical revelation of origins could be satisfactorily worked out (actually, of course, as we shall see later, such a harmonization is quite impossible), most Christians would be quite content to leave the subject to these scientists to work out to their own satisfaction, and would not concern themselves about it.

---

[6] *The Intelligent Man's Guide to Science,* Vol. 1 (New York; Basic Books, Inc., 1960), p. 80.

[7] *Historical Geology* (New York; Wiley, 2nd ed., 1961), p. 9.

But, as Huxley pointed out in the quotation cited above, evolution has intruded itself into every area of life. It has become the basic undergirding philosophy of all the social sciences, the humanities, and even the study of religion itself, so that it is impossible to ignore its implications.

Consider, for example, the fields of psychology and psychiatry, the study of the human mind and mental health. These disciplines bear tremendous potential for influence over future social orders and disciplines, whether for good or ill. It is therefore quite disturbing when such a man as Henry W. Brosin, Chairman of the Department of Psychiatry at the University of Pittsburgh, in his contribution to the 1959 Centennial Convocation at the University of Chicago, says:

> "It is appropriate for psychiatrists and other students of mental disorders to pay homage to the work of Charles Robert Darwin and the theory of evolution, for without this work it is difficult to imagine what the state of our discipline would be like."[8]

Or consider the field of sociology, with its multiplicity of concerns in every area of human life and activity – marriage, the home, social institutions, crime, economic controls, cultural units, etc. An authoritative summary of this field states:

> "The master idea, which animated alike the initiator of sociology (i.e., Auguste Comte) and his chief continuator (Herbert Spencer), was that of evolution. . . . Independently of the writings of both Comte and Spencer, there proceeded during the 19th century, under the influence of the evolutionary concept, a thoroughgoing transformation of older studies like History, Law and Political Economy; and the creation of new ones like Anthropology, Social Psychology, Comparative Religion, Criminology, Social Geography. It is from these sources that have sprung the main body of writings, investigation, research, that today can properly be called sociological."[9]

---

[8] "Evolution and Understanding Diseases of the Mind," in *The Evolution of Man* (Sol Tax, Ed., University of Chicago Press, 1960), p. 373.

[9] Victor Branford: "Sociology," article in *Encyclopedia Britannica,* Vol. 20, (University of Chicago Press, 1949), p. 912. Branford was, at the time of writing, Chairman of the Council of the British Sociological Society.

Space is not available for discussion of the influence of the evolutionary philosophy upon such fields as philosophy, literature, economics, and other disciplines in the humanities and social sciences, but this influence has been profound.

> "*The Origin of Species* has had more influence on Western culture than any other book of modern times. It was not only a great biological treatise, closely reasoned and revolutionary, but it carried significant implications for philosophy, religion, sociology, and history. Evolution is the greatest single unifying principle in all biology."[10]

I think it is highly important to emphasize, however, that *all* of the anti-Christian systems of modern times have found their quasi-scientific basis in the supposed scientific fact of evolution. This has been true for communism and for the various varieties of socialism, for modern militarism, and even for the anti-Christian aspects of modern capitalism and colonialism. In fact, it seems that the advocates of any doctrine or system overtly or covertly espousing covetousness or selfishness in any form have appealed to evolutionary science as warrant for their opinions.

> "Our own generation has lived to see the inevitable result of evolutionary teaching—the result that Sedgwick foresaw as soon as he had read the *Origin*. Mussolini's attitude was completely dominated by evolution. In public utterances, he repeatedly used the Darwinian catchwords while he mocked at perpetual peace, lest it hinder the evolutionary process. In Germany, it was the same. Adolf Hitler's mind was captivated by evolutionary teaching—probably since the time he was a boy. Evolutionary ideas—quite undisguised—lie at the basis of all that is worst in *Mein Kampf*—and in his public speeches."[11]

The racism and militarism of Hitler and Mussolini were in large measure built upon the philosophical base established in the 19th century by Friedrich Nietzche and Ernst Haeckel, both of whom were rabid promulgators of Darwinism among human societies.

And this was no less true of communism, although the application may have been different. It is well known that Marx

---

[10] C. L. Prosser: "The *Origin* after a Century: Prospects for the Future." *American Scientist,* Vol. 47, December, 1959, p. 536.

[11] R. E. D. Clarke: *Darwin: Before and After,* (Paternoster Press: London, 1948), p. 115.

wished to dedicate his book, *Das Kapital,* to Darwin. Socialistic writers, such as Herbert Spencer, Enrico Ferri, Morris Hilquitt, Paul Blanshard, and innumerable others, as well as communists such as Marx, Engels, and the whole spectrum of modern Soviet writers, have continually referred their doctrines of class struggle, economic determinism, atheistic materialism, etc., to their proper foundation in Darwinism.

> "Like the right-wing politicians, the revolutionary socialists saw in evolution a way of stifling their consciences. The new doctrine justified men in struggling for their rights—even though the struggle involved a denial of the Christian virtues."[12]

In fact, it may safely be affirmed that nearly all writers of "liberal" persuasion, whatever their particular "ism" may be, start from evolutionary presuppositions. Liberalism espouses a doctrine of "progress," based on man's innate potentialities for development, a doctrine which implicitly denies the fact of man's Fall and basic depravity. It is all too little realized how thoroughly and completely this false philosophy of evolutionary progress has permeated human culture, even here in supposedly Christian America.

A good example of the far-reaching influence of this idea in America is the fact that practically the entire structure of modern public school education is centered around this theme. This fact is so obvious and so common to universal observation and experience that it needs no documentation.

The evolutionary origin of the universe, of life and of man is taught as scientific fact even to elementary school children in probably most public schools, at least by implication. The Christian and Biblical record of origins is usually ignored, sometimes allegorized or even ridiculed. Such concepts as Creation, the Fall, the Curse, Sin, Redemption, etc. – which really are the most important and basic facts of science and history – are taboo. Such patronizing references to "religion" as are allowed at all in the public schools are given in a context of comparative religions, of "world communities," of "brotherhood," of "social progress," and the like.

---

[12] *Ibid,* p. 113.

This is a remarkable phenomenon in a nation founded largely on Christianity and the Bible. Undoubtedly, many factors have contributed to this "devolution," but it is highly probable that the introduction of the 19th-century doctrines of evolutionary optimism is back of most of it.

The individual exerting the greatest single influence over the development of modern educational theory and practice in this country is, beyond any doubt, John Dewey. A recent writer has evaluated him thus:

> "But John Dewey as a great apostle of political liberalism exerted an all-pervading influence on the American scene. He more than any other has thoroughly saturated political liberalism with ethical relativism."[13]

Dewey's contributions to psychology, philosophy, sociology, and other fields are numerous and important, and all based on Darwinian evolution. But his greatest influence has been as the architect of the modern system of progressive education, through the stimulus of his Schools of Education at Columbia University and the University of Chicago. The moral and spiritual damage inflicted on the youth of America for two generations by this system has been incalculable. And it is profoundly significant that Dewey was completely evolutionistic in his entire approach. As his biographer says:

> "The starting-point of his system of thought is biological: he sees man as an organism in an environment, remaking as well as made. Things are to be understood through their origins and functions, without the intrusion of supernatural considerations."[14]

Besides the school, the greatest institution affecting American life, as in other Christian countries, would certainly be the church. And the greatest tragedy of all is that the Christian church, in such large measure, has capitulated to the doctrine of evolution. The rise of religious modernism in the 19th century followed the scientific propaganda of Darwinism, and was in large measure based on it. Theologians were persuaded that

[13] N. R. Van Til: "Can a Theological Conservative Be a Political Liberal?" *Torch and Trumpet,* Vol. XII, November 1962, p. 17.

[14] Will Durant: Article "John Dewey," in *Encyclopedia Britannica,* Vol. VII, 1956, p. 297.

science required acceptance of the geological ages and the evolutionary origin of the various species, including man. Various schemes for harmonizing the creation account in Genesis with evolution were developed, but none was really valid, and the necessary outcome was finally the complete rejection of the records of Genesis as nothing but myth and legends.

The defection of the liberal wing of Protestantism is too well recognized even to require documentation. The great denominations – Methodist, Presbyterian, Episcopalian, Baptist, and others – have in large measure long since accepted evolution and adapted their Biblical exegesis and theology thereto in such wise as necessary. It is not yet so generally known that the supposedly more conservative bodies – Roman Catholicism and evangelicalism – have also in recent years capitulated to a large extent.

Official Catholic doctrine, as enunciated in the encyclical, *Humani Generis,* by Pope Pius XII in 1950, permits the belief and teaching of evolution in the church, provided that the moral and spiritual character of man is still recognized as a divine creation. Although many Catholics have been reluctant to accept evolution, the scholastic leadership of the church is largely subservient now to this philosophy. Gottfried O. Lang, Professor of Anthropology at the Catholic University in Washington, is a typical spokesman. In a Darwinian Centennial symposium, he says:

> "Is then evolution fact or fancy? From what has been said, the question seems to be a spurious one. There are the facts of change, the facts of variability of organic forms, and there is the undeniable fact of the great antiquity of some simple forms. Unless one wants to go back to the days of the 'special creationists' these facts must be interpreted in some such way as was indicated above. But it must be kept in mind that this is *an* explanation. There may be others forthcoming. But the weight of evidence at this time seems to favor such an explanation."[15]

[15] "Human Organic Evolution, Fact or Fancy?" in *Symposium on Evolution* (ed. B. O. Boelen, Duquesne University, 1959), p. 54.

The conservative wing in Protestantism has been the last stronghold of anti-evolutionary scholarship, as well as the source of much anti-evolutionary emphasis that could hardly be denoted as "scholarly." But in recent years, even a large segment of this group has yielded to evolution. However worthy may have been the motives of those leaders involved in this trend, presumably hoping to reach a larger number of educated men with an intellectually acceptable interpretation of Scripture, the trend is undeniable and immensely dangerous.

Illustrating this trend is the following statement by J. Frank Cassel, recent president of the American Scientific Affiliation, an organization of over one thousand evangelical scientists committed to the belief that the Bible is the Word of God:

> "Thus, in 15 years we have seen develop in A.S.A. a spectrum of belief in evolution that would have shocked all of us at the inception of our organization. Many still reserve judgment but few, I believe, are able to meet Dr. Mixter's challenge of, 'Show me a better explanation.' Some may see in this developing view the demise of our organization, but it seems to me that we only now are ready to move into the field of real potential contribution—that in releasing Truth from the restrictions we have been prone to place upon it, we can really view it in the true fullness which the Christian perspective gives us."[16]

Views such as this have become quite common among the "neo-evangelicals." Various terms are used to describe, and somewhat to obfuscate, these views — terms such as "threshold evolution," "progressive creation," and the like. These all imply that God has been gradually developing organic species slowly over long ages, and thus in effect are merely variations of theistic evolution.

This "spectrum of belief in evolution" which Dr. Cassel noted as increasingly predominant in the American Scientific Affiliation, may also be easily recognized in many areas of erstwhile evangelical conviction. No purpose would be served to cite names of individuals, colleges, seminaries, publications, etc.,

---

[16] "The Evolution of Evangelical Thinking on Evolution," *Journal of the American Scientific Affiliation,* Vol. 11, No. 4, December, 1959, p. 27.

which reflect this trend, but its existence is quite apparent to all who are interested enough to investigate. How rapidly and how universally this acceptance of evolution will permeate the thinking of Biblical Christians remains to be seen.

But to the extent that it does penetrate, one can anticipate a rapid subsequent yielding on other points of theology and Biblical interpretation as well. Acceptance of evolution is logically followed by the rejection of a high theory of Biblical inspiration, then by rejection of the doctrine of the fall and the curse, and finally by the rejection of the substitutionary atonement. This historic sequence has been repeated again and again in churches, schools, denominations, periodicals, and many other Christian enterprises and organizations over the past one hundred years. In a very real sense, the theological capitulation to evolution has been the forerunner and the basis of the development of modernism in religion. This is true not only of the old pre-War modernism or religious optimism but also of the so-called Neo-orthodoxy, which has so swept the post-War religious world. In his brief but incisive review of the development of Barthianism, Charles Caldwell Ryrie, Dean of the Graduate School, Dallas Theological Seminary, says:

> "Out of Darwin and his so-called scientific method grew the higher criticism of the 19th century.... What were the results of this teaching of liberalism? There was a high and false estimate of the ability of human nature. It promoted the illusion that the kingdom of God was capable of being fulfilled in history, for man's ability could bring this to pass. There came with it the abandonment of the distinctive and exclusive character of the message of the Gospel and the loss of the uniqueness of Christianity; and secularization of life and thought can also be traced at least in part to the teaching of liberalism in the last century."[17]

Dr. Ryrie proceeds to show how the leaders of Neo-orthodoxy (Barth, Brunner, Niebuhr, *et al.*), while professing a return to Biblical theology, continue to reject the accounts of Creation, the Fall, etc., as historic facts, and thus continue to build their religious system upon an evolutionary philosophy as its foundation:

---

[17] *Neo-orthodoxy* (Chicago: Moody Press, 1956), p. 18.

"Thus the Genesis account of creation and the fall is rejected as history—as most of us understand history. Science, the Barthians say, has delivered us from having to believe the Genesis stories, and through this scientific deliverance, we are supposed to be able to see the real meaning of the accounts."[18]

In summary, we have seen that the evolutionary philosophy has all but universally captivated modern thought. The concept of evolution undergirds present-day sociology, psychology, philosophy, economics, educational theory, and religion, as well as the natural sciences, both physical and biological. Its impact upon modern life may not always have been obvious, but it is certainly tremendous. And to a large extent, that influence has been deleterious. Evolution is at the foundation of communism, Fascism, Freudianism, social Darwinism, behaviourism, Kinseyism, materialism, atheism and, in the religious world, modernism and Neo-orthodoxy.

Jesus said: "A good tree cannot bring forth corrupt fruit" (Matthew 7:18). In view of the bitter fruit yielded by the evolutionary system over the past hundred years, a closer look at the nature of the tree itself is well warranted today.

And it is encouraging to note that a great many capable and educated men today are becoming willing to take a much closer and more critical look at the theory of evolution than has been true heretofore. It is still true, of course, that most intellectuals accept evolution. In a television panel discussion just prior to the famous Darwinian Centennial Celebration at the University of Chicago, a number of most interesting observations were made by the participants. These participants included the following: (1) Sol Tax, Professor of Anthropology at the University of Chicago, and Chairman of the Darwinian Celebration; (2) Sir Julian Huxley, perhaps the world's most famous evolutionary biologist, grandson of Thomas Huxley, and first Director-General of UNESCO; (3) Sir Charles Darwin, grandson of *the* Charles Darwin, and world-famous physical scientist and population expert; (4) Harlow Shapley, Professor of Astronomy at Harvard, prolific writer on astronomy and cosmic evolution, and frequent sponsor and joiner of Communist front organizations; and (5)

---

[18] *Ibid*, p. 51.

Adlai Stevenson, now United States representative at the United Nations. A more influencial cross-section of leaders of modern thought could hardly be imagined. Several statements taken more or less at random from the discussion will be noted.

Huxley started off by asserting:

> "The first point to make about Darwin's theory is that it is no longer a theory, but a fact. No serious scientist would deny the fact that evolution has occurred, just as he would not deny the fact that the earth goes around the sun."[19]

Adlai Stevenson, after noting that he had been writing editorials criticizing William Jennings Bryan at the Scopes trial when he was only twenty-three years old, raised the question:

> "I wonder if there is a parallel between the resistance to teaching about evolution in our schools and the resistance to teaching about Marxism and communism in this country. Until very recently, we practically equated teaching about Marxism (which is absolutely imperative for understanding the Russians) with advocating subversion. I think we are much more enlightened now."[20]

Concerning the implications of Darwinism for religion, Huxley stressed:

> "Darwinism removed the whole idea of God as the creator of organisms from the sphere of rational discussion. Darwin pointed out that no supernatural designer was needed; since natural selection could account for any known form of life, there was no room for a supernatural agency in its evolution. . . . There was no sudden moment during evolutionary history when 'spirit' was instilled into life, any more than there was a single moment when it was instilled into you. . . . I think we can dismiss entirely all idea of a supernatural overriding mind being responsible for the evolutionary process."[21]

To which sentiments, Sir Charles Darwin interjected his hearty agreement. Dr. Shapley then entered the discussion to point out that it was not necessary to berate religion so, since most modern religionists accepted evolution in almost the full context demanded by Huxley:

---

[19] "At Random: A Television Preview," in *Issues in Evolution* (Vol. III of *Evolution after Darwin,* Sol Tax, Ed., Univ. of Chicago Press, 1960), p. 41.

[20] *Ibid,* p. 44.

[21] *Ibid,* p. 45.

> "You spoke of their parting. But there are many kinds of religions. I have had much contact with the liberal clergy of America in the last two or three years; and they accept evolution, without objecting to it or worrying about it."[22]

In spite of the unanimous agreement among evolutionists that all intelligent people agree with them, however, there are many who do not. Furthermore, the writer is convinced, from having discussed the subject with hundreds of people, that the main reason most educated people believe in evolution is simply because they have been told that most educated people believe in evolution! Very rarely is such a person able to do more than repeat a few stock "evidences for evolution," and almost never has he given any really serious consideration to the question of their real implications.

For when a man will really examine critically the nature of these evidences, he will find that very serious difficulties and contradictions abound in them. Probably more people are doing just that today than at any time since the rise of Darwinian evolution. For example, it is quite significant that more than a thousand scientists who subscribe to the doctrine of an inspired and inerrant Bible have joined the American Scientific Affiliation. It is not known what proportion of these believe in evolution (the A.S.A. leadership at present seems to have a preponderance of theistic evolutionists, and presumably a large part of the rank-and-file share similar views), but there are undoubtedly a great many of them who *do not*. The writer has the privilege of being a Fellow of this organization and there are quite a number of other A.S.A. Fellows who are anti-evolutionists, all of them qualified and recognized scientists, with the doctorate degree or its equivalent.

Furthermore, only a very small percentage of those scientists who are non-evolutionists are in the A.S.A. The writer has spent a total of twenty-two years on the faculties of five universities (Rice, Minnesota, Southwestern Louisiana, Southern Illinois, and Virginia Polytechnic Institute). At each of these places, there were a number of men who were not only conserv-

---

[22] *Ibid.*, p. 46.

ative Christians but who did not believe in evolution. At V.P.I., for example, the writer knows personally more than twenty-five faculty members who fit this description. None of these schools is a Christian school in any sense of the word; four are state-supported universities and the fifth, Rice, is private, and its tone may be described by noting that Julian Huxley was on the faculty there for the first four years of its existence!

The writer has been publishing books and articles defending Biblical Christianity and Creationism during most of those twenty years, and one interesting and rewarding aspect of these efforts has been the large number of communications received from scientists in every field and from all parts of the world, most of them quite sympathetic, if not in full agreement. Opportunities to speak in churches and meetings of many different kinds have been quite numerous, and almost always there seem to be scientists or other educated people in the group who express hearty agreement. At a recent meeting (September 1962) of the Houston Geological Society, the largest local geological society in the world, the writer was invited to speak on the subject: "Biblical Castastrophism and Geology." After the meeting, a substantial number of geologists came up to express deep interest, and some to express full agreement.

Consequently, it is decidedly unnecessary to agree with the constant refrain from Huxley and others that "all intelligent persons agree that evolution is a fact." It may be true that not many anti-evolutionists are vocal about their convictions, whatever the reasons for this may be. But there are many of them.

One reason for the apparent dearth of anti-evolutionary sentiment is that the major scientific publishing houses and periodicals are completely and exclusively under the control of leaders who are evolutionists. If anyone questions this, let him try to get a serious scientific article or book published refuting evolution! Or even a Letter to the Editor! The only outlet for such literature seems to be through conservative or private media.

Similarly, it is almost an impossibility for a convinced creationist to obtain or to retain an influential position on a university faculty in the various disciplines now dominated by the evolution concept, such as anthropology, geology, biology, psychology, and psychiatry. The writer has known some men personally, and has heard of others, who were refused graduate degrees in geology, for example, primarily on the basis of their rejection of Lyellian uniformitarianism and Darwinian evolutionism.

The university scientists who reject evolution referred to in the foregoing are obviously, therefore, mostly in the physical sciences – chemistry, physics, engineering, mathematics, etc. Most of the biologists, geologists, and psychologists who reject evolution are working in private industry or in governmental positions.

In spite of the overwhelming monopoly that evolutionists have developed over educational and communications media, however, there does exist a tremendous reservoir of intelligent anti-evolutionary Christian conviction in this and other countries. And if well-written scholarly literature of this nature could somehow be channeled to the great body of educated "men of good will," who are inclined to believe in evolution simply because of a brainwashing process to which they have been subjected ever since entering the public schools but whose minds are not closed to new considerations, then there is no doubt that a much greater body of anti-evolutionary sentiment could be quickly developed.

And this should not be an impossible task. It is not too difficult to demonstrate that the entire concept of evolution is not only anti-Biblical but also utterly unscientific. That is, it is not difficult to demonstrate this to one who is willing to think in terms of Biblical and Christian presuppositions. And for those whose presuppositions, often unconsciously, are anti-Christian, it would at least be salutary for them to encounter the fact that their methods and conclusions have necessarily been colored by their premises.

Chapter II

The Case Against Evolution

In this chapter and the next we shall summarize the evidence against evolution by showing, first, that there is no evidence of evolution occurring at present and, second, that there is no evidence that evolution has occurred in the past. In doing this, it is necessary to start with the Biblical record. Particularly in the past, prior to human historical records, it is manifestly impossible to prove scientifically whether evolution took place or not. In the nature of the case, the history of the earth and its inhabitants cannot be subject to scientific experimentation; the events are non-reproducible and, therefore, not legitimately subject to analysis by means of the so-called "scientific method."

One must, therefore, either start with the assumption that God is the Creator and the Author of history, or else with the assumption that there is no God and that the history of the earth and the universe is to be explained without him. The way one approaches the study of this history must necessarily depend upon the assumption with which he starts. If one more or less arbitrarily ignores God in developing such a history, even though he may not deliberately intend to exclude the possibility of God, in effect he is making the second assumption and is taking the approach of atheism. For it should be plainly emphasized that, if God *does* exist and if he *is* the Creator and Sustainer of history, then it is foolhardy to attempt to understand history apart from his revealed Word. In other words, the only way we can know with *certainty* the time of creation, the order of creation, the meaning of creation, the methods of creation, and anything else concerned with pre-historical events, is for God to tell us these things. He was there and we were not. Therefore, in every case, we believe that the only legitimate method of reasoning in this sphere is the *deductive* method. One starts with either one assumption or the other and then develops his system and his conclusions. He *cannot* use the inductive method, attempting to build up a historical record on the basis of bits of evidence he may be able to find in the present world. In doing this, he is in reality using the deductive method but starting with the atheistic assumption that God has *not* already spoken concerning these things.

We, therefore, must simply start with the assumption that God exists and is the Creator and Sustainer of this universe. Consequently we must acknowledge that God *can* reveal himself if he so wills and that it is not possible for us really to understand *anything* (since our very minds have been created by him) unless he does so. The Bible claims in numberless ways to be this revelation, and has validated its claims in equally innumerable ways. Therefore, in any historical or scientific argumentation, here is where we start.

With respect to the possibility of evolution occurring in the present or in the past, we must first of all define clearly what is meant by evolution. Evolution does *not* simply mean *change*. This is important, because the evidence cited by most writers in favor of their claim that evolution is a *fact* is simply evidence of change. But true evolution is a certain kind of change.

Once again, we shall let evolution's chief present-day spokesman and protagonist, Sir Julian Huxley, settle this particular question:

> "Evolution is a one-way process, irreversible in time, producing apparent novelties and greater variety, and leading to higher degrees of organization, more differentiated, more complex, but at the same time more integrated."[1]

---

[1] *Ibid.*, p. 44. See also footnote 4, chapter 1.

This statement was intended to include both inorganic and organic evolution, and to comprehend the whole of the physical and biological universes. That is, everything in the universe has been developed by this process of evolution, of development, of progress, of higher and higher levels of organization and complexity.

With this definition in mind, we come to examine the question of whether there is any evidence that such a process is *now* taking place in the world. And the answer, both Scripturally and scientifically, is, unequivocally, *no!*

As far as the Bible is concerned, this process of organization, of increasing complexity, of development, of integration, is simply the process of creation. And, according to Scripture, creation is no longer taking place.

> "Thus the heavens and the earth were *finished,* and all the host of them. And on the seventh day God *ended* his work which he had made; and he rested on the seventh day from all his work which he had made. And God blessed the seventh day and sanctified it: because that in it he had *rested* from all his work which God created and made" (Genesis 2:1-3).

> "For in six days the Lord made heaven and earth, the sea, and *all that in them is,* and *rested* the seventh day: wherefore the Lord blessed the sabbath day, and hallowed it" (Exodus 20:11).

> "It is a sign between me and the children of Israel for ever: for in six days the Lord made heaven and earth, and on the seventh day he *rested,* and was refreshed" (Exodus 31:17).

> "By the word of the Lord were the heavens made; and all the host of them by the breath of his mouth. For he spake, and it was *done;* he commanded, and it *stood fast*" (Psalm 33:6, 9).

> "Thou, even thou, art Lord alone; thou *hast made* heaven, the heaven of heavens, with all their host, the earth, and all things that are therein, the seas and all that is therein, and thou *preservest* them all" (Nehemiah 9:6).

> "By the word of God the heavens were *of old,* and the earth *standing* out of the water and in the water" (II Peter 3:5).

> "The works were *finished* from the foundation of the world" (Hebrews 4:3).
>
> "For he that is entered into his rest, he also hath *ceased* from his own works, as God did from his" (Hebrews 4:10).

These passages of Scripture, in both Old and New Testaments, make it plain that the work of creation was terminated at the end of the six days. God is now *preserving* everything he had created in the six days, but he is no longer *creating* anything.

God has, therefore, told us plainly in his Word that nothing is now being either created or destroyed, and we are, therefore, not surprised when, as we study the laws of nature, we find that the most basic, the most universal, the best-proved, law of all science is the law of *Conservation!*

Actually, there are many so-called conservation laws of science. Conservation of mass, conservation of linear momentum, conservation of electric charge, conservation of angular momentum, and conservation of energy are the most important. And without doubt the one truly universal conservation law is that of energy conservation, especially when broadened to include possible mass-energy conversions.

Energy, defined as "capacity for doing work," actually includes everything in the physical universe. Because of mass-energy equivalence, all forms of matter are, in a very real sense, merely forms of energy. Energy may also appear as mechanical, electrical, electro-magnetic, chemical, light, heat, sound, and other types of energy.

> "The First Law of Thermodynamics is merely another name for the Law of Conservation of Energy.... This law states that energy can be transformed in various ways, but can neither be created nor destroyed."[2]

All processes in the universe – physical, geological, biological, etc. – involve transformations of energy. It is not too much to say that the whole of physical reality is merely the outworking of

[2] A. R. Ubbelohde: *Man and Energy* (New York; George Braziller, Inc., 1955), p. 149. Dr. Ubbelohde is Professor of Thermodynamics at the Imperial College of Science and Technology in the University of London, and also Dean of the Faculty of Science at Queen's University in Belfast.

the energies of the universe. And all of this is fundamentally described and controlled by the law of energy conservation, which states that mass-energy is neither being created nor destroyed. And this is precisely what the Biblical revelation has told us!

Furthermore, it ought to be evident that this universal law squarely contradicts, and therefore disproves, the evolutionary hypothesis, which maintains that "creation" – that is, increasing organization and integration and development – is continually taking place in the present.

And if the first law of thermodynamics disproves evolution, what could one say about the second law of thermodynamics! The second law, equally universal and also proved beyond any scientific doubt whatever, states that in all energy transformations there is a tendency for some of the energy to be transformed into non-reversible heat energy. That is, the availability of the energy of the system or process for the performance of work is reduced. It "runs down" or "wears out." The term *entropy* is used as a measure of the amount of energy thus depleted from the system, and the second law states, therefore, that the entropy of a closed system can never decrease, but rather always tends to increase.

The second law of thermodynamics was originally developed by Carnot, Clausius, and Kelvin, starting from work on the engineering problems of steam engines. In its early forms, it was developed at about the same time as Darwin's publication of *Origin of Species.* However, its broader implications were only gradually becoming understood by the end of the 19th century. Even today, it is obvious that most people, especially most evolutionists, have very little understanding of the tremendous implications of the second law.

> "Understanding of the law has continued to grow since the time of Clausius and Kelvin.... In its most modern forms, the Second Law is considered to have an extremely wide range of validity. It is a remarkable illustration of the ranging power of the human intellect that a principle first detected in connection with the clumsy puffing of the early steam engines

should be found to apply to the whole world, and possibly even to the whole cosmic Universe."[3]

The physicist R. B. Lindsay, Dean of the Brown University Graduate School, says concerning the universal importance of the two laws of thermodynamics:

> "Thermodynamics is a physical theory of great generality impinging on practically every phase of human experience. It may be called the description of the behaviour of matter in equilibrium and of its changes from one equilibrium state to another. Thermodynamics operates with two master concepts or constructs and two great principles. The concepts are *energy* and *entropy,* and the principles are the so-called first and second laws of thermodynamics. . . ."[4]

There is no real question, either, that the two laws apply to biological systems as well as physical systems. In fact, practically all evolutionary biologists today reject vitalism in biology, insisting that all biological processes are really only physico-chemical processes, with no "vital force" or "vital energy" involved. It thus follows that these physico-chemical processes in living systems must conform to the two laws of thermodynamics. The significance of this becomes clear when the second law is defined in most general terms. As implied above, its implications are far wider than contained in the tendency for processes to produce irrecoverable heat energy. The thermodynamic application is in reality only a special case of a universal tendency for everything to become more "probable" – that is, more disorganized, more "random." The Princeton biologist, Harold Blum, applying this fact to biological systems, makes this quite clear:

> "A major consequence of the second law of thermodynamics is that all real processes go toward a condition of greater probability. The probability function generally used in thermodynamics is *entropy*.
> . . . The second law of thermodynamics says that left to itself any isolated system will go toward greater entropy, which also means toward greater randomization and greater likelihood."[5]

---

[3] *Ibid.,* p. 146.

[4] "Entropy Consumption and Values in Physical Science," *American Scientist,* V. 47, September, 1959, p. 376.

[5] "Perspectives in Evolution," *American Scientist,* V. 43, October, 1955, p. 595.

It would hardly be possible to conceive of two more completely opposite principles than this principle of entropy increase and the principle of evolution. Each is precisely the converse of the other. As Huxley defined it, evolution involves a continual *increase* of order, of organization, of size, of complexity. The entropy principle involves a continual *decrease* of order, of organization, of size, of complexity. It seems axiomatic that both cannot possibly be true. But there is no question whatever that the second law of thermodynamics is true!

Of course, it is quite possible for entropy to *decrease* in an *open* system. In fact, every instance of a local increase in organization – the growth of a child, the development of a crystal, the raising of a building – is an example of the influx of an excess of "energy" or "information" into the particular open system, so that its innate tendency toward decay is *temporarily* offset. But that child, or crystal, or building, or *anything else* will *eventually* start to grow old or wear out or decay. Even the temporary, supposedly natural growth of an organism is really to be attributed ultimately to the creation and maintenance by God of a marvelous mechanism of reproduction and sustenance.

And remember that evolution, in the minds of its proponents, is not a localized phenomenon anyway, but rather a universal law, explaining alike the development of species in biology, elements in chemistry, and suns in astronomy! As Huxley insists: "The whole of reality is evolution."

It is hard to believe that the leaders in evolutionary thought, not to mention their hosts of uncritical followers, have ever really confronted this gross contradiction between their *theory* of evolution (which they protest overmuch to be a "fact") and the second *law* of thermodynamics. For example, the great Darwinian Centennial Celebration at the University of Chicago in 1959, which brought together the acknowledged leaders in this field from all over the world, and which produced many original papers and much discussion, apparently never even recognized the existence of this problem. In the three volumes of papers and discussions emanating from this conference, it is almost impossible to find any mention of questions of this sort at all.

Although of course some may have been missed, a fairly careful search indicates that only two of the writers[6] in this Symposium refer to it, and these only briefly and cursorily.

And until this fundamental contradiction is thoroughly cleared up and harmonized, creationists are abundantly justified in insisting that evolution as a universal principle is not only unproved but statistically almost impossible! The second law of thermodynamics plainly and relentlessly insists that there is a universal tendency toward decay and disorder, not growth and development. This is true on the cosmic scale and, even though it may temporarily be negated on a small scale by local increases in order resulting from external influences, even these are only temporary and will eventually decay.

But this is not at all surprising to the Christian, for this is what is taught in the Word of God. Not only has God told us that he has *finished* his creation, and is now *preserving* it, so that nothing further is being created nor is anything being destroyed, but he has also told us that there is everywhere in the world a tendency toward decay and death. Everything, left to itself, tends to grow old and to run down and finally to die.

> "Of old hast thou laid the foundation of the earth: and the heavens are the work of thy hands. They shall perish, but thou shalt endure: yea, all of them shall wax old like a garment; as a vesture shalt thou change them, and they shall be changed" (Psalm 102:25, 26).

> "Lift up your eyes to the heavens, and look upon the earth beneath: for the heavens shall vanish away like smoke, and the earth shall wax old like a garment and they that dwell therein shall die in like manner: but my salvation shall be forever, and my righteousness shall not be abolished" (Isaiah 51:6).

> "For the creation was made subject to vanity. . . . For we know

[6] Hans Gaffron: "The Origin of Life," in *The Evolution of Life* (Vol. I of *Evolution after Darwin,* Sol Tax, Ed., University of Chicago Press, 1960), p. 40.

Alfred E. Emerson: "The Evolution of Adaptation in Living Systems." *Ibid.*, p. 312.

> that the whole creation groaneth and travaileth in pain together until now" (Romans 8:20, 22).
>
> "For all flesh is as grass, and all the glory of man is as the flower of grass. The grass withereth and the flower thereof falleth away" (I Peter 1:24).
>
> "All go unto one place; all are of the dust, and all turn to dust again" (Ecclesiastes 3:20).
>
> "Heaven and earth shall pass away" (Matthew 24:35).

Not only does the Bible tell us the *fact* of decay in the creation, but it also gives us the explanation for it, something which thermodynamics has not been able to do. The universal validity of the second law of thermodynamics is demonstrated, but no one knows *why* it is true. It is strictly an empirical law, which has always been found to be true wherever it could be tested, but for which there is no known natural explanation.

But the Biblical explanation is that it is involved in the curse of God upon this world and its whole system, because of Adam's sin. At the end of the six days of creation, the Scripture says that "God saw everything that he had made, and, behold, it was very good" (Genesis 1:31). If there be any doubt as to what is meant by this, it is clarified by the description of conditions in the new earth, which will be created by God after this present system has passed away. In Revelation 21:4, it is promised that there will then be no more (1) sorrow, (2) pain, (3) crying, or (4) death. That all of these things are associated with the curse on the present world is evident from the parallel statement in Revelation 22:3, which says that in the new earth, "there shall be no more curse." And it is also evident from the actual description of the curse as given in Genesis 3:17, as we shall note below.

Therefore, we conclude that the Bible teaches that, originally, there was no disorder, no decay, no aging process, no suffering, and above all, no death, in the world when the creation was completed. All was *"very good."*

But, "by one man, sin entered the world, and death by sin" (Romans 5:12). Eve sinned, and Adam sinned, the essence of both acts being rejection of the word of God. Eve listened to the words of Satan, and Adam harkened to the words of his wife, and both thereby explained away the word of God, and then flagrantly refused to obey his word. Fellowship with their Creator was broken, and the perfect order of God's creation and purpose was disturbed by the entrance of disorder and rebellion into the world. Since Adam had been designated master of all the earth and everything in it (Genesis 1:28), the curse likewise comprehended everything under Adam's dominion.

According to the Biblical record, the curse is as follows:

> "Cursed is the ground [or *earth,* which is an alternate rendering of the Hebrew] for thy sake; in sorrow shalt thou eat of it all the days of thy life; thorns also and thistles shall it bring forth to thee; and thou shalt eat the herb of the field; in the sweat of thy face shalt thou eat bread, till thou return unto the ground; for out of it wast thou taken: for dust thou art and unto dust shalt thou return" (Genesis 3:17-19).

As noted above in connection with the eventual removal of the curse from the earth, there are four main elements in it: (1) sorrow; (2) pain – symbolized by the thorns and thistles; (3) crying, that is, the groaning and struggle and intense effort necessary to wrest a living from a reluctant earth, all intimated by the sweat; and (4) death, when the highly organized protein and other structures of the body will finally break down and decay and eventually return to the basic elements – the "dust of the earth" from which it was made.

All of this can be summarized in terms of a great principle of decay and disorder in the earth. Adam was originally commissioned to "subdue" the earth and to exercise dominion over it, but now he and his descendants must reckon with an earth which resists his efforts. Only by continual effort and overcoming all manner of difficulties can order be maintained or increased. In the struggle there will be encountered much pain and sorrow, both of which manifest an inharmonious environment, external and internal. And ultimately, regardless of all the sorrow and sweating and pain overcome in "eating of" the earth, the earth will finally be victorious and will regain her "dust."

Can there be any doubt that here, and here only, we have the real explanation for the relentless increase of entropy in the world? As the physicist, R. B. Lindsay, has said, concerning the second law of thermodynamics:

> "All experience points to the fact that every living organism eventually dies. This is a process in which the highly developed order of the organism is reduced to a random and disorderly collection of molecules. We are reminded that we are 'dust' and to 'dust' we ultimately return."[7]

The exact physiological mechanism which is responsible for aging and death of an animal has never been fully determined, and this is in fact an active area of modern research. As Howard Curtis, of the Brookhaven National Laboratory, says:

> "Everyone realizes that he will undergo adverse changes, with the passage of time, which will eventually lead to death in one form or another, and accepts this as inevitable. It is difficult to think of a biological process of more interest to most adults, and yet through the years the explanations for this phenomenon have mostly been couched in vague generalities. Even today gerontologists cannot agree upon a definition of aging."[8]

After discussing various suggested causes for aging, Curtis presents strong modern evidence that the main cause is found in somatic mutations. These are sudden changes in the structure of the somatic cells (as distinct from the germ cells which transmit genetic character from parent to offspring), brought about by radiation or other mutagens affecting the organs and general body cell structure of the animal. He says:

> "Certainly the vast majority of mutations must be deleterious, so if the organs of older animals contain appreciable numbers of cells which are carrying mutations, it is a virtual certainty that the organs are functioning less efficiently than they otherwise would."[9]

These somatic mutations have no effect upon evolution, because, as is now well established, acquired characters cannot be inherited. However, similar mutations occur in the germ cells, and these can be and are transmitted to descendants, as discussed later. These genetic mutations must have a similarly deleterious effect upon the species as a whole, just as the somatic

---

[7] "Entropy Consumption and Values in Physical Science," *American Scientist,* Vol. 47, September, 1959, p. 384.

[8] "Biological Mechanisms Underlying the Aging Process," *Science,* Vol. 141, August 23, 1963, p. 686.

[9] *Ibid,* p. 688.

mutations seem clearly to lead to the aging and death of the individual. However, the germ cells are much better protected from factors causing mutation than are the somatic cells. As Curtis says:

> "It is suggested that the mutation rates for somatic cells are very much higher than the rates for gametic cells, and that this circumstance insures the death of the individual and the survival of the species."[10]

But even many species eventually decay and die, with the accumulated effects of generations of mutations and a hostile environment, including especially the presence of man. The effects of the Fall and the Curse are both worldwide and agelong, and there is really no other satisfactory way of accounting for the fact that the "whole creation groaneth and travaileth in pain together until now." In a recent Phi Beta Kappa address, the noted anthropologist, Loren Eiseley, has said:

> "As one gropes amid all this attic dust it becomes ever more apparent that some lethal factor, some arsenical poison seems to lurk behind the pleasant show of the natural order or even the most enticing cultural edifices that man has been able to erect."[11]

But, of course, the Word of God not only reveals the cause of the universal decay, but also reveals that it will not last forever. The so-called "heat death" anticipated by scientists as the ultimate fate of the universe, when all free energy has been utilized, and converted into non-available heat energy, will never be reached.

> "Because the creation itself also shall be delivered from the bondage of corruption [that is, *decay*] into the glorious liberty of the children of God" (Romans 8:21).

> "For the earnest expectation of the creation waiteth for the manifestation [that is, revealing] of the sons of God" (Romans 8:19).

> "Nevertheless we, according to his promise, look for new heavens and a new earth, wherein dwelleth righteousness" (II Peter 3:13).

---

[10] *Ibid,* p. 694.

[11] "Man, the Lethal Factor," *American Scientist,* Vol. 51, March, 1963, p. 72.

This, perhaps, is not the best place for a gospel message, but all this is nevertheless an outworking of the gospel. The revealing of the redeemed children of God, the deliverance of the creation, the new heavens and the new earth, are all made possible by the tremendous fact of Jesus Christ. God, in Christ, has redeemed the world from sin and death by himself dying for the sins of the whole world (I John 2:2) and his bodily resurrection from the grave. At present, he is "taking out a people for his name" (Acts 15:14), regenerating those whom he calls and who "believe on his name" (John 1:12, 13), making them, through the indwelling of his Holy Spirit, the "sons of God" (Romans 8:14), who will be openly manifested as such when Christ "shall appear" (I John 3:2, 3) – that is, when Christ comes again to this world at the end of this age.

But until he comes, the whole creation continues in the bondage of decay. Physical systems, left to themselves, run down and stop; biological organisms grow old and die; societies isolated from uplifting influences deteriorate and vanish away; individuals, who reject or neglect the regenerating influences of the gospel or its by-products, soon drift downward morally and spiritually, as well as physically, and finally die.

> "Then when lust hath conceived, it bringeth forth sin: and sin, when it is finished, bringeth forth death" (James 1:15).

And this is all so absolutely contrary to the whole concept and philosophy of evolution that one could scarcely conceive more diametrically opposed systems. The two systems are alike in only one respect, in that both involve continual *change*. But one is a change *up*, the other a change *down*. One is development, the other deterioration; one growth, the other decay.

Here we encounter in our study of this subject a very remarkable phenomenon. This fact of *change*, which is both Biblically and scientifically observed to be a universal implication of the second law of thermodynamics, has been appropriated by evolutionists as the evidential basis for their theory.

No one would question that change *occurs*. New varieties of various species are developed, by means of various types of biological mechanisms. In most cases, however, these changes are quite definitely within narrow limits. All the varieties of dogs remain inter-fertile and are still dogs, for example. Within all recorded human experience, it is highly questionable whether evolutionists can point with assurance to more than *this* kind of change occurring. The Mendelian laws of heredity provide for much variation on the basis of the outworking of the genetic factors present in the chromosome structure of the germ cells of each species. But such variation (or, depending on definitions, perhaps sometimes speciation) *always* has definite limits.

This is exactly the situation that would be expected on the basis of the Genesis account of creation. Nothing in the account indicates how many original "species" there were, or what constitutes a "species." However, it does clearly indicate that there were meant to be definite limits to the possible biological changes that might take place. The only biological unit identified therein is called a *kind,* and at least ten times in the first chapter of Genesis is it mentioned that the various types of living creatures were to bring forth "after their kind." This states, quite plainly, that there were to be definite limits to possible biological change, perhaps, by implication, these limits being those of inter-fertility. But within those limits, it can surely be inferred that variation and speciation are possible. An interesting comment on the unsettled state of the "species problem" in modern biological research is given in a recent article by two Stanford biologists:

> "The term *species* should be retained only in its original, less restrictive sense of 'kind.' There seems to be no reason why quantitative methods should not be used to study phenetic relationships (those based on similarity rather than imagined phylogeny) at what we now loosely call the species level."[12]

But changes of this type have nothing much to do with what evolutionists consider to be true evolution. Mere reshuffling of genetic factors already present is not evolution. This process corresponds analogically to energy transformations in a physical system, with nothing really gained or added – just the form changed. Rather, some permanent and hereditable change must occur of an entirely different type than those potentially present already. Such changes are called "mutations," and are brought

---

[12] Paul R. Ehrlich and Richard W. Holm: "Patterns and Populations," *Science,* Vol. 137, August 31, 1962, p. 655.

about by a definite and sudden change in one or more genes in the germ cell. Bonner says:

> "[Mutation] is really the factor of fundamental importance. Since mutation means a chemical change in the gene structure, all progressive advancements must ultimately be by mutation, and all that can be done by recombination is to shuffle what is given by mutation. Gene mutation provides the raw material for evolution, and recombination sets this material out in different ways so that selection may be furthered by being provided with a whole series of possible arrangements."[13]

That true mutations do occur, and that these are hereditable, and may result in permanent *change* in the species, no creationist need question in the least. But the important point is that these changes are fully in line with the universal law of deterioration; in fact, that is exactly what such changes amount to.

For a mutation is essentially a sudden and apparently random change in the genetic structure of the germ cell, brought about by penetration of the cell by radiation, a mutagenic chemical or some other *disorganizing* agent. The effect is analogous to what would happen to, say, a television picture tube if a bomb were exploded inside it. There would be a change, all right, but it would, in all probability, not be an improvement! (This might depend on one's point of view with respect to television programs, however.)

> "Mutations and mutation rates have been studied in a wide variety of experimental plants and animals, and in man. There is one general result that clearly emerges: almost all mutations are harmful. The degree of harm ranges from mutant genes that kill their carrier, to those that cause only minor impairment. Even if we didn't have a great deal of data on this point, we could still be quite sure on theoretical grounds that mutations would usually be detrimental. For a mutation is a random change of a highly organized, reasonably smoothly functioning living body. A random change in the highly integrated system of chemical processes which constitute life is almost certain to impair it."[14]

---

[13] *The Ideas of Biology* (New York; Harper and Brothers, 1962), p. 64.

[14] James F. Crow: "Genetic Effects of Radiation," *Bulletin of the Atomic Scientists*, Vol. 14, January, 1958, pp. 19-20.

Evolutionists are hard pressed to find any actually observed mutations, as distinguished from mere recombinations of genetic factors, which are helpful in the struggle for existence. Occasionally a rare mutation, such as bacterial resistance to penicillin, may accidentally result in improved ability to cope with a changed environment. And it is these occasional helpful mutations occurring in artificially changed environments which are actually proposed by evolutionists as the biological mechanism accounting for the entire development of all living organisms through geologic time! The hypothetical process of natural selection is supposed to act on these occasional mutations in such a way as to preserve those rare ones which are beneficial. Actually, the more complex an organism, the less chance there is of a mutation being beneficial in any environment. This is a principle of such generality as to have status fully as valid as that of most other physical "laws," or putting it another way, the more complex a structure, the less probable it is that a random change will increase its complexity. Therefore, the mutation concept of evolution seems about as logical as to say that, if a man travels south ninety-nine miles, then north one mile, then south ninety-nine miles, then north one mile, and so on, he will reach the North Pole before he reaches the South Pole!

Mutations really, therefore, offer a perfect illustration of the second law of thermodynamics, which says that the natural tendency of all change is to create a greater degree of disorder and randomness. This would mean that the over-all direction of change of a biological "kind" would be deteriorative rather than developmental. This is evident not only in the case of present genetic changes, but also in those evidences that have been cited in favor of past evolutionary changes. For example, the evidence of vestigial organs is often cited as an argument for evolution. But it is immediately evident that the loss of organs through disuse is an illustration of deterioration.

Similarly, paleontology reveals that practically every type of living creature in the present world has ancestors in the fossil record which are larger than their present-day descendants. One thinks, for example, of the mammoths, the cave bears, sabertooth tigers, giant bisons, the dinosaurs, the giant beavers, cockroaches, rhinos, and even giant men! The evolutionary increase in size and complexity supposedly revealed by the fossil record apparently breaks down in the transition from the hypothetical sequences of the geologic past to the actual creatures of the present! And, as we shall see later, these hypothetical phylogenies of the fossil record can be interpreted in an alternate manner which supports, rather than contradicts, the second law of thermodynamics.

Before leaving this subject, it would be well to note a recent theory that has attempted to sidestep the problems posed by the second law of thermodynamics.

> "A recent suggestion is that for the Universe considered as a whole the law of entropy increase is brought to a standstill by the 'continuous creation' of matter. The hypothesis of 'continuous creation' has in fact been introduced in the attempt to neutralize the law of entropy-trend on the cosmic scale."[15]

This theory of the "steady-state" universe has been widely publicized and popularized in recent years. A group of British astronomers, especially Hoyle and Bondi, have been the leaders in the promulgation of this strange hypothesis. It is *mis*called the "continuous creation" theory, since it does not postulate that God is still creating anything. In fact, it is thoroughly atheistic, since it assumes that the universe never had a beginning at all, and will never have an end. It arbitrarily decides that the universe should always be essentially the same, at any point of time or space. In order to eliminate the profound difficulty imposed upon such a theory by the second law of thermodynamics, which rigidly interpreted would require *both* a beginning and an end of the universe as observed, it allows for the continual *evolution* (not "creation") of matter out of nothing!

It must be clearly recognized that there cannot possibly be any observational or experimental basis in support of such a notion. It is simply required by the assumption of a "steady-state" universe, with neither beginning nor end. Its proponents argue that this assumption is so reasonable that it warrants the

---

[15] A. R. Ubbelohde: *Man and Energy*, p. 177.

otherwise absurd idea of continual evolution of matter out of nothing.

This is merely a striking confirmation of the earlier assertion that a man's presuppositions will determine how he handles the scientific data. But even many uniformitarian scientists are appalled at the presumption of Hoyle and his colleagues in promoting such a theory as this in the name of science. In a recent review of several new books bearing on this theme, G. C. McVittie, Head of the University of Illinois Astronomy Department, says:

> "The temptation to substitute logic for observation is peculiarly hard to resist in astronomy. This is because astronomical data are hard to come by, and the data rapidly diminish in number and accuracy as the objects we observe recede from the earth. . . . Nevertheless, the fact that data may be scarce and inaccurate is no reason for failing to use them as our main guides in the formulation of theory. . . . Once upon a time, British science was sometimes criticized for being too empirical. During the past 30 years a number of *a priori* theories of cosmology, of which the steady-state theory is one, have completely reversed the trend, which is a curious and unexpected development."[16]

We conclude this section, then, by reiterating that the revealed Word of God, supported completely by all true science, teaches that the evolutionary principle, as applied to present processes and events, is not only not valid but is essentially impossible. The basic processes at the present time are those of *conservation* and *deterioration,* not innovation and development.

[16] "Rationalism versus Empiricism in Cosmology," *Science,* Vol. 133, April 21, 1961, p. 1236.

Chapter III

The Testimony of Geologic History

As far as *present-day* biological change is concerned, therefore, we insist that there is no evidence whatever that any real evolutionary changes are now taking place. Genetic variations, certainly in at least the overwhelming majority of instances, are within rigidly fixed limits, so that the basic species remain essentially unchanged. When change occurs outside of these limits, as a result of mutations of some kind, then again in the overwhelming preponderance of cases, the change is either harmful or, at best, neutral to the creature experiencing it.

These facts are in perfect accord with the two universal laws of thermodynamics, which describe a *universal condition of quantitative stability and qualitative deterioration.* At the very best, therefore, either quantitative or qualitative evolution must be accomplished by means of some sort of mechanism which, locally and temporarily, may be able to supersede the effects of the laws of thermodynamics. Natural selection is supposed by evolutionists to be the needed mechanism. A modern leader in this field says:

> "The general picture of how evolution works is now clear. The basic raw material is the mutant gene. Among these mutants most will be deleterious, but a minority will be beneficial. These few will be retained by what Muller has called the sieve of natural selection. As the British statistician R. A. Fischer has said, natural selection is a 'mechanism for generating an exceedingly high level of improbability.' It is Maxwell's famous demon superimposed on the random process of muta-

> tion. Despite the clarity and simplicity of the general idea, the details are difficult and obscure."[1]

The last statement above is strikingly descriptive of the entire theory of evolution. The idea is simple and powerfully persuasive to the natural mind, but the details of evidence supporting it become increasingly obscure the more closely they are examined. The "beneficial minority" of mutations which can supposedly be preserved by natural selection, for example, is vanishingly small, and the almost infinite accumulation of beneficial mutations that would be required for the true evolution of even a single major kind of animal surely requires natural selection to be a remarkable type of mechanism, one which can truly generate an "*exceedingly* high level of improbability." Maxwell's demon, indeed! It is much easier to suppose that the very idea of evolution was generated by this ubiquitous demon!

If, then, there is no evidence for true evolution occurring in the present, the only way in which the fact of evolution could be demonstrated would be to show that it had occurred in the past, throughout geologic time. It is everywhere admitted that there has been no more evolution in historic times than there is occurring in the present. In fact, the most ancient written records of plant and animal life reveal no significant changes of a truly evolutionary nature to have occurred at all. Walter E. Lammerts has recently reminded us of this:

> "As K. Patau has shown, even mutations having a one per cent survival advantage increase in frequency from 0.01 to 0.1 per cent of the population only after 900,230 generations. Another 100,511 generations are needed to increase the frequency to 100 per cent. Certainly the time needed for natural selection to effect a change in a large population is enormous even geologically speaking. That is why Sir Charles Lyell's concept of slow change by presently acting causes is so necessary for any concept of general evolution."[2]

---

[1] James F. Crow: "Ionizing Radiation and Evolution," *Scientific American,* Vol. 201, September, 1959, p. 142.

[2] "Growing Doubts: Is Evolutionary Theory Valid?", *Christianity Today,* Vol. VI, September 14, 1962, p. 4.

Since neither present nor past human observations record any evidence of true evolution, it is necessary to buttress the theory by claiming that evolution occurred in *pre*-historic times. In effect, the evolutionist says: "Although we can't prove that evolution has occurred within historic times, it must have occurred in the past in order for the present state of the biological world to have been attained. Therefore, it must be occurring in the present, and anyone who doubts the *fact* of evolution is hopelessly ignorant!"

The only evidence (apart from divine revelation, which the evolutionist refuses to accept) concerning pre-historic life on the earth is that which can be deduced from the fossil remains of creatures now buried in the rocks of the earth's crust. These fossiliferous deposits are interpreted to show a gradual evolution of the earth and its inhabitants over long ages, and this is considered the real core of the evidence supporting the theory of evolution. As the Yale geologist, Carl O. Dunbar, says:

> "Although the comparative study of living animals and plants may give very convincing circumstantial evidence, fossils provide the only historical, documentary evidence that life has evolved from simpler to more and more complex forms."[3]

But in what way do fossils of dead animals provide evidence for evolution? Since they were deposited in most cases prior to human historical observations and records, it is obviously impossible to know for certain just how and when they lived and were buried. In order to interpret their testimony, one must start from some premise concerning their significance and then attempt to deduce a theory which can explain the data on the basis of his premise. If he cannot do this, then he must try some other premise and the resulting theory. Even if he does hit upon a satisfactory theory, which seems to explain the data he still cannot be certain that it is right, since it may be possible to find several theories which can correlate all the data to at least some degree.

The almost universally promoted theory for interpreting the fossils is summarized by Dunbar as follows:

---

[3] *Historical Geology* (New York: Wiley, 2nd Ed., 1961), p. 47.

> "Since fossils record life from age to age, they show the course life has taken in its gradual development. The facts that the oldest rocks bear only extinct types of relatively small and simple kinds of life, and that more and more complex types appear in successive ages, show that there has been a gradual development or unfolding of life on the earth."[4]

This superficially seems very convincing and, indeed, is so convincing that it is really, as we have seen, the very foundation of the theory of evolution, which, in turn, has been appropriated as the philosophical basis of nearly all modern disciplines of human knowledge.

But at least two important questions must be satisfactorily answered before it can legitimately be concluded that the theory of evolution is the best explanation for the fossil record. One question is: "Are the ages of the rocks determinable independently of the theory of evolution which is supposed to be deduced from their fossil contents?" The other is: "Is the theory of evolution the *only* theory which can satisfactorily explain the fossil data?" Both of these questions must be answered in the affirmative if we should be expected to accept the fossils as real *proof* of evolution. But as a matter of fact, both questions must really be answered in the negative.

The problem of determining the age of a given rock formation is very important to this whole issue. How is it decided which rocks are old and which are young and, in general, how do we determine the entire chronology of geologic time? Again, we shall let Professor Dunbar explain:

> "Inasmuch as life has evolved gradually, changing from age to age, the rocks of each geologic age bear distinctive types of fossils unlike those of any other age. Conversely, each kind of fossil is an *index* or *guide fossil* to some definite geologic time. . . . Fossils thus make it possible to recognize rocks of the same age in different parts of the Earth and in this way to correlate events and work out the history of the Earth as a whole. They furnish us with a chronology, 'on which events are arranged like pearls on a string.' "[5]

Examination of this statement makes it immediately obvious that there is a subtle example of circular reasoning here. Rocks

---

[4] *Ibid.*
[5] *Ibid.*, pp. 47-48.

are dated by the fossils they contain, rocks containing simple fossils being considered old and vice versa. This amounts simply to assuming as a prior fact that evolution is known to have occurred throughout geologic time. Then, the resulting geologic column, with its fossil series, is said to be the main, and indeed the only, proof that evolution has occurred.

This is such an important point that we shall call in other authorities as confirming witnesses. Cornell geologists, O. D. von Engeln and Kenneth E. Caster, state:

> "The geologist utilizes knowledge of organic evolution as preserved in the fossil record, to identify and correlate the lithic records of ancient times."[6]

E. M. Spieker, of Ohio State, emphasizes that the geologic time-scale is based predominantly on the paleontological evidence (that is, on the fossil sequences postulated by evolution) rather than on any physical evidence (such as the physicochemical nature of the rocks, or their relative position in terms of vertical succession, etc.):

> "And what essentially is this actual time-scale . . . on what criteria does it rest? When all is winnowed out, and the grain reclaimed from the chaff, it is certain that the grain in the product is mainly the paleontologic record and highly likely that the physical evidence is the chaff."[7]

One of the most prominent European paleontologists has said:

> "The only chronometric scale applicable in geologic history for the stratigraphic classification of rocks and for dating geologic events exactly is furnished by the fossils. Owing to the irreversibility of evolution, they offer an unambiguous time scale for relative age determinations and for world-wide correlations of rocks."[8]

In spite of this frank recognition of the pre-eminent importance of the fossils in dating rock formations, the obvious circle of reasoning involved in this process is rarely admitted, at least

---

[6] *Geology* (New York: McGraw-Hill, 1952), p. 417.

[7] "Mountain-Building Chronology and Nature of Geologic Time-Scale," *Bulletin, American Assoc. of Petroleum Geologists,* Vol. 40, August, 1956, p. 1803.

[8] O. H. Schindewolf: "Comments on Some Stratigraphic Terms," *American Journal of Science,* Vol. 255, June, 1957, p. 394.

in print.[9] One exception was R. H. Rastall, of Cambridge University, who said:

> "It cannot be denied that from a strictly philosophical standpoint geologists are here arguing in a circle. The succession of organisms has been determined by a study of their remains buried in the rocks, and the relative ages of the rocks are determined by the remains of organisms that they contain."[10]

Now of course these men, as well as other geologists and paleontologists, would hasten to insist that, even though the time scale is built upon the basis of an assumed evolution, the resulting system is so consistent and so universally verified that this assumption is fully validated. That is, the fossils are always found in the same order, no matter in what part of the world they are discovered, and always the order is from simple to complex. Rocks buried lowest have the simpler fossils and those nearer the surface have the more complex fossils. The "geologic column" is the same everywhere.

But this is simply not so, in spite of the evolutionists' wishful thinking which would like for it to be so. There are great numbers of exceptions and contradictions to this generalization. As a matter of fact, the geologic column really exists only in the minds of the historical geologists, since it has been built up by superposition of deposits from various parts of the world.

> "If a pile were to be made by using the greatest thickness of sedimentary beds of each geologic age, it would be at least 100 miles high. . . . It is, of course, impossible to have even a considerable fraction of this at one place. The Grand Canyon of the Colorado, for example, is only one mile deep. . . . By application of the principle of superposition, lithologic identification, recognition of unconformities, and reference to fossil successions, both the thick and thin masses are correlated with other beds at other sites. Thus, there is established, in detail, the stratigraphic succession for all the geologic ages."[11]

---

[9] A number of geologists have acknowledged this verbally, off the record.

[10] Article "Geology," in Encyclopedia Britannica, 1956, Vol. 10, p. 168 (University of Chicago Press).

[11] O. D. von Engeln and Kenneth E. Caster: *Geology*, pp. 417-418.

Neither is much of the geologic column present at any one site necessarily continuous. Fossils are almost always found in sedimentary rocks, and below whatever sedimentaries are found in a given location there will always be found at the bottom crystalline rocks of the so-called "basement complex." The latter presumably is remnant from that period in the earth's history before the formation of sedimentary rocks began. It is significant that literally any rock system in the entire geologic column may be found lying directly on the basement complex and that any combination of rock systems may be found above it, at any given location.

> "Further, how many geologists have pondered the fact that lying on the crystalline basement are found from place to place not merely Cambrian, but rocks of all ages?"[12]

And, similarly, any series of rock systems may be found above the bottom, and there need be no difference in appearance, except for the fossils they contain.

> "An unconformity separating the oldest Pre-Cambrian from the latest Pleistocene may have the same physical appearance as one between the latest Pleistocene and the middle Pleistocene. The fossils of the strata bounding an unconformity are the only indicators of time-value, and these are not always decisive for determination within narrow limits."[13]

An unconformity is supposedly a boundary between two rock formations of widely different ages, supposedly caused by erosion during those ages. These often are found with perfectly parallel bedding and every other appearance of immediate succession of deposition instead of long ages intervening. They are then called, variously, "disconformities," "paraconformities," or even "deceptive conformities."

Far more serious than this is the fact that it is common to find supposedly "ancient" rock formations resting in essential conformity on supposedly "young" formations. This is exactly contrary to the requirements of evolution, which would necessitate that the oldest rocks should be at the bottom. Neverthe-

---

[12] E. M. Spieker: "Mountain-Building Chronology and Nature of Geologic Time-Scale," p. 1805.

[13] W. H. Twenhofel: *Principles of Sedimentation* (2nd Ed., New York, Mc-Graw-Hill, 1950), p. 562.

less, this anomalous condition is quite common. Carl O. Dunbar admits these conditions in the following words:

> "In disturbed areas, of course, the normal succession may be locally inverted, as in the lower limb of an overturned fold, or it may be interrupted or duplicated, by faults, but such abnormalities will betray themselves in evidences of disturbance and in an unnatural sequence of fossils."[14]

Although sometimes there may be evidences of physical disturbance (leading to faulting and folding) in these "upside-down" areas, it is quite often true that they can only be revealed by an "unnatural sequence of fossils," which means that the fossils are not found in the order pre-supposed by their evolutionary relationships. Walter E. Lammerts comments:

> "The actual percentage of areas showing this progressive order from the simple to the complex is surprisingly small. Indeed formations with very complex forms of life are often found resting directly on the basic granites. Furthermore, I have in my own files a list of over 500 cases that attest to a reverse order, that is, simple forms of life resting on top of more advanced types."[15]

In order to account for these numerous exceptions to the supposed universal order of evolutionary development as revealed in the fossiliferous rocks, theory has to be piled on top of theory. Thus, the missing ages indicated by a disconformity are explained by a supposed regional uplift and period of erosion. An inverted order of fossils is explained by a regional uplift followed by a horizontal thrust fault followed by a period of erosion. And so forth. One is reminded of Occam's Razor, the principle that cautions against any unnecessary multiplication of hypotheses to explain a given set of phenomena.

In any case, it becomes obvious that the theory of evolution does not really provide a very simple and satisfactory framework for the correlation of the data of paleontology. It is emphatically clear that evolution is assumed in building up the geologic time scale and that, even so, there are so many problems involved that subsidiary theories continually have to be appended to it in order to explain the exceptions and

---

[14] *Historical Geology* (New York, McGraw-Hill, 1961), p. 9.

[15] "Growing Doubts: Is Evolutionary Theory Valid?", p. 4.

contradictions.[16] The charge of circular reasoning which has been lodged against the critically important paleontological evidence of evolution is not simply to be laughed off or ignored as evolutionists too commonly attempt to do. It quite plainly involves the presupposition of evolution, with numerous involved deductions based on that premise. It is *not,* therefore, valid to offer this presupposition and these deductions as *proof* of evolution, and especially in view of the tremendously important fact that there is no real evidence of *present* evolution, and of the even more significant fact that the two universal laws of thermodynamics plainly imply universal stability or deterioration rather than evolution!

The first of the two questions asked concerning this main proof of evolution, namely its logical independence of presuppositions involving its own proof, must therefore be answered emphatically in the negative. But even if it did not involve circular reasoning and even if it did present a fully consistent explanation of the fossil record, the second question would still have to be faced. Is evolution the only, or even the best, possible explanation of the fossils? And the answer is, most certainly, that it is not! The Biblical revelation of early terrestrial history, together with the solid scientific foundation of the first and second laws of thermodynamics, lead to a much more satisfying explanation of the fossil record than does the theory of evolution.

The Biblical framework involves three major facts of history, each of tremendous importance with respect to the scientific study of data bearing on these problems. These facts are of such obvious significance that to ignore them means that one is arbitrarily rejecting even the possibility that God could have given a genuine revelation of beginnings in His Book of Beginnings. The three facts are: (1) a real Creation; (2) the

---

[16] Space limitations necessarily restrict discussion of the problems involved in the evolutionary interpretation of the paleontological data. For a considerably fuller discussion, see *The Genesis Flood,* by John C. Whitcomb, Jr., and Henry M. Morris (Nutley, N. J.: Presbyterian & Reformed Publishing Co., 1961), pp. 130-211.

Fall of man and resultant Curse on the earth; and (3) the universal Deluge in the days of Noah.

The first two have already been discussed in part. According to the Bible, God created all things in heaven and earth, including all living kinds of animals, as well as man, in the six-day period of creation. Following this period of creation, He rested. Thus, no true creation is now taking place in the world, and this revelation is confirmed by the great principle of mass and energy conservation.

Now this can only mean that, since nothing in the world has been created since the end of the creation period, everything must *then* have been created by means of processes which are no longer in operation and which we therefore cannot study by any of the means or methods of science. We are limited exclusively to divine revelation as to the date of creation, the duration of creation, the method of creation, and every other question concerning the creation. And a very important fact to recognize is that true creation *necessarily* involves creation of an "appearance of age." It is impossible to imagine a genuine creation of anything without that entity having an appearance of age at the instant of its creation. It would always be possible to imagine some sort of evolutionary history for such an entity, no matter how simple it might be, even though it had just been created.

This is seen most clearly in the record of the creation of Adam and Eve. According to the record, Adam was created as a mature man, formed by God out of the elements of the physical earth. He was not created first as an embryo or a baby, and then allowed to develop. Similarly, Eve was created directly out of Adam. In like manner, everything was created as a fully developed, perfectly functioning whole. Soil was created for the plants to grow in; chemical molecules and compounds were created; light from the sun and stars and moon was seen on the earth at the instant of their creation; and so on. Thus, everything in the earth *must* have had an appearance of age, if there had been any true creation at all. The earth and universe constitute a great clock which was originally wound up by God, in a manner and at a time which can only be known, if at all, by means of divine revelation. The "apparent age" at which the "clock" was originally set may have been anything that pleased him. In any case, when the creation was finished, God judged it all to be "very good" – perfectly functioning and fully harmonious, with nothing incomplete or out of order, and then God "rested." And this primeval condition continued until "sin entered into the world."

The possibility of creation of apparent age is recognized by even such a doctrinaire evolutionist as George Simpson, Professor of Vertebrate Paleontology at Harvard University, who says:

> "We cannot disprove the postulate that the universe was created one second ago, complete with all our apparent memories of our own earlier days, or that it was not created in 4004 B.C., with all the apparent record of earlier billions of years. But that would not make sense, and we must pretend, at least, that both we and the universe are sane."[17]

Simpson is obviously caricaturing the problem and, since he is an avowed disbeliever in any divine purpose in the universe, the concept of "creation" of any kind to him would not "make sense." Others would say that the concept of apparent age involves the Creator in some kind of deception and, therefore, they reject it out of solicitude for the divine honor. But, as we have pointed out above, to say that God could not create anything with apparent age is tantamount to saying nothing could be created and, therefore, is essentially the same position as the atheism of Simpson. In fact, rather than honoring God's truthfulness by rejecting any supposed "deception" on his part in creating apparent age, such men in reality are charging him with falsehood, since they deny the truth of his revealed Word concerning the creation. We insist as emphatically as we know how that the doctrine of creation of apparent age does not in the remotest degree involve a divine deception, but is rather inherent in the very nature of creation. Further, God in grace has even revealed much concerning the true age of the crea-

---

[17] "The History of Life," in *The Evolution of Life* (Sol Tax, Ed., University of Chicago Press, 1960), p. 175.

tion, in His written Word, but men have simply refused to accept it.

The second great revealed fact of earth history is that of the fall of man, followed by God's divine curse on the whole creation. The effects of the curse, manifested particularly in the universal tendency toward decay and disorder and death in the world, have been discussed somewhat already. The second law of thermodynamics has been seen to approximate a scientific statement of the effects of the curse.

For our present purposes, the point to be noticed is that, in the fossil record, there is an abundance of testimony to the effect that decay, disorder, and death have existed in the world during all the geologic ages represented by the fossil-bearing rocks. The very fact of fossil animals demonstrates the fact of death, and there is also much geologic evidence of disease, of physical catastrophes, of suffering, of struggle – in short, of a pre-historic world which was "groaning and travailing together in pain." Even the very concept of evolution itself, especially as furthered by natural selection, involves a struggle for existence, with the strong exterminating the weak. It is certainly difficult to imagine that this was the state of things at the end of the creation period when God, who is Love, saw everything that he had made and pronounced it to be "very good."

The Biblical framework, therefore, requires that we categorically reject the fossil record as a record of the history of the *development* of life on the earth. It cannot possibly be ascribed to the period and events recorded in the first chapter of Genesis, during which God was creating the heavens and the earth and everything in them.

And the same is the testimony derived from the two laws of thermodynamics. There is certainly no indication that the sedimentary rocks of the earth's crust, with their fossil contents, were laid down under conditions and at a time when the two laws were not in existence. Such a notion would certainly be completely contrary to the doctrine of uniformitarianism which supposedly governs the interpretation of the geologic records. Now, since the bringing into existence of matter and energy and order – in a state of low entropy (that is, of high order and high energy availability) – essentially requires a process or processes of creation, and since such processes are precisely opposite to those of stability and deterioration postulated by the first and second laws of thermodynamics, it therefore follows that the period of true creation (even call it true evolution, if you will) could not possibly have been the same period as the period represented by the deposition of the fossil-bearing rocks. They simply cannot scientifically be regarded as a record of the evolution (or "creation," as preferred) of higher and higher forms of life on the earth.

But then what *do* they represent? They must have been laid down *after* the introduction of the present order of things into the universe and deposited under the action of the present physical laws which now control the behavior of nature. (Uniformitarians should not object to this statement!) This means that they must have been deposited *after* God finished creating all things, since the law of conservation of mass and energy was in operation when they were laid down. It also means that they must have been deposited *after* God pronounced the curse on the creation, since they were laid down while the second law of thermodynamics was in operation. More directly to the point, they could only have been deposited after *death* entered into the world, which means after man had sinned! Therefore, it is both scientific and Scriptural to insist that the fossil deposits of the earth's crust must have all been brought into place at some time or times after the creation and fall of man.

But these deposits are so extensive and so thick, spread all over the earth's crust, sometimes to a depth measured in miles, that it is quite impossible that they could have been formed by the ordinary processes of deposition that are taking place at present. They were undoubtedly formed under the operation of the same basic physical *laws* that now exist (and this is true uniformitarianism), but they could not have been formed by geologic processes acting at the same *rates* as at present. Rather these processes (sedimentation, erosion, volcanism, tectonism, radioactivity, glaciation, etc.) must have operated at greatly augmented rates and over greatly enlarged areas. In short, the old geologic doctrine of catastrophism, which has been stigmatized ever since the days of Lyell and Darwin, must be revived if there is to be any hope of a scientific accounting for the facts of the fossil record.

Once this is recognized, and the doctrinaire uniformitarianism of the past one hundred years rejected as completely unable to correlate all the facts of the record, then it will be seen that a much more satisfactory explanation of the fossil record can be developed than is possible on the basis of evolution. Two Canadian geologists contrast catastrophism and uniformitarianism as follows:

> "One of the aids to the interpretation of sedimentary rocks is the principle of uniformitarianism. This principle states that the processes we see at work on the crust of the earth today are sufficient to account for all the events of the past that have formed the crust. In other words, 'the present is the key to the past.' When the science of geology was young and the great age of the earth unknown, geologists believed that the features of the crust were formed by a series of catastrophes."[18]

This type of uniformitarianism has been extremely important in geologic interpretation for over a century.

> "This is the great underlying principle of modern geology and is known as the principle of *uniformitarianism*. . . . Without the principle of uniformitarianism there could hardly be a science of geology that was more than pure description."[19]

With true uniformitarianism, based on the strict application of the two laws of thermodynamics and other basic physical laws, we have no quarrel whatever. For if these universal laws were given full consideration in the development of a geologic history, it would soon be recognized that evolution is practically impossible statistically and that, therefore, the geologic data must be explained in terms of creation and subsequent deterioration. But this perfectly legitimate and proper application of the principle of uniformity has been ignored in favor of the entirely unwarranted assumption that secondary processes

---

[18] Thomas H. Clark and Colin W. Stearn: *The Geological Evolution of North America,* (New York: Ronald Press, 1960), pp. 5, 6.

[19] William D. Thornbury: *Principles of Geomorphology* (New York: Wiley, 1954), pp. 16, 17.

such as radiogenic accumulations, erosion and deposition must have always been occurring, not in accordance with the same physical *laws* as at present, but rather at the same *rates* as at present! For this assumption, there is not the slightest warrant whatever, except that it yields the tremendous expanse of geologic time that is necessary to give even a semblance of plausibility to the evolutionary hypothesis.

As a matter of fact, this assumption does not at any point provide a satisfactory explanation for the geologic data. For example, the most recent geological epoch before the present one is called the Pleistocene. By all rights, if the standard geologic time table is at all valid, the record of this epoch ought to be the plainest and easiest to interpret in uniformitarian terms. But, on the other hand, the Pleistocene has been interpreted in terms of a geologic catastrophe of first magnitude, namely, as the great Ice Age, or perhaps series of ice ages! And so inadequate is the principle of uniformity, for explaining the onset, oscillations, and decline of the great complex of continental ice sheets supposedly characteristic of this period, that theories by the carload have been proposed, each in turn rejected for one reason or another! As Loren Eiseley has recently observed:

> "Even on a more dramatic scale no one to date has been quite able satisfactorily to account for that series of rhythmic and overwhelming catastrophes which we call the Ice Age."[20]

Other types of geologic deposits fare little better in terms of a really consistent application of geologic uniformitarianism. Practically all fossils are found in sedimentary rocks, especially in shales and limestones, and most of these were presumably laid down in relatively shallow water, such as might be found along the continental shelves. But little success has been attained in relating these sedimentary rocks to actual sedimentary environments of deposition, such as are now actually observed in the present. A prominent marine geologist, Francis P. Shepard, has said:

> "Most sedimentary rocks are believed to have been deposited

[20] "Man, the Lethal Factor," *American Scientist,* Vol. 51, March, 1963, p. 74.

in the seas of the past. One of the primary purposes in geological investigations has been to interpret the conditions under which these ancient sediments were deposited. One of the obvious places to look for guidance in these interpretations is in the deposits of the present. It is, therefore, rather surprising to find how little attention geologists had paid to these recent marine sediments until very recent years."[21]

Similarly, the great expanses of volcanic terrains in the Pacific Northwest, the Canadian Shield, the Indian plateaus, and many other places, have to be explained in terms of great systems of volcanic vents and fissures which are completely incommensurate with any type of volcanic activity ever observed by man in modern times. The tremendous earth movements implied by the great faults and folds in the earth's crust and even by the obviously recent uplifts of most of the world's great mountain regions certainly have no present-day parallel. In fact, wherever one looks in the deposits of the crust, he finds phenomena that cannot possibly be explained in terms of present-day rates of geologic processes. The inadequacies of geological uniformitarianism to account for the fossiliferous rocks have been rather thoroughly discussed and documented[22] and need not be discussed in more detail here.

One further aspect of this particular problem should be mentioned, however, and that is the matter of the fossils themselves. Remember that the fossils in the rocks provide the very means of dating the rocks and that the paleontologic series thus constructed is the only real proof of evolution. Consider also that these rocks are supposed to have been laid down by means of the slow operation of geologic processes occurring at present.

And then meditate upon the remarkable fact that, for the most part, the fossils simply *must* have been laid down under sudden and probably catastrophic conditions or else they would never have been preserved as fossils at all! Even such a consistent evolutionary uniformitarian geologist as Dunbar recognizes that practically all fossils must have been formed by floods or other catastrophes.

[21] "Marine Sediments," *Science,* Vol. 130, July 17, 1959, p. 141.
[22] Whitcomb and Morris, *op. cit.,* pp. 130-169.

"A carcass left exposed after death is almost sure to be torn apart or devoured by carnivores or other scavengers, and if it escapes these larger enemies, bacteria insure the decay of all but the hard parts, and even they crumble to dust after a few years if exposed to the weather. If buried under moist sediment or standing water, however, weathering is prevented, decay is greatly reduced, and scavengers cannot disturb the remains. For these reasons burial soon after death is the most important condition favoring preservation. . . . Water-borne sediments are so much more widely distributed than all other kinds, that they include the great majority of all fossils. Flooded streams drown and bury their victims in the shifting channel sands or in the muds of the valley floor."[23]

Other catastrophes such as falls of volcanic ash can account for some large concentrations of fossils. In fact, it could be said that with only insignificant exceptions, all of the fossils must have been deposited by some kind of catastrophe or else they would not have been preserved at all. Normal rates of sedimentation, etc., as postulated by the uniformity principle, are meaningless as far as the fossil record is concerned.

And even the occasional destructive flood or volcanic eruption that occurs in modern times cannot be taken as typical of the cause of the most important fossil deposits. Some of these have no modern parallel at all. The fantastic deposits of hundreds, possibly millions, of mammoths and other animals in the mucks of the Arctic provide an example. The great "fossil graveyards" found in many parts of the world, sometimes containing millions of fish, sometimes hordes of dinosaurs or other animals, sometimes a heterogeneous mixture of animals of all kinds, all testify that present-day rates and phenomena cannot possibly account for them.

Modern paleontologists are beginning to be more realistic than they once were with regard to the necessity of at least some degree of catastrophism in the interpretation of the fossil record. Norman D. Newell, of Columbia University and the American Museum of Natural History, has recently commented:

"Yet the fossil record of past life is not a simple chronology of uniformly evolving organisms. The record is prevailingly one of erratic, often abrupt changes in environment, varying rates of

---

[23] *Op. cit.*, pp. 35-36, 39.

> evolution, extermination and repopulation. Dissimilar biotas replace one another in a kind of relay. Mass extinction, rapid migration and consequent disruption of biological equilibrium on both a local and a worldwide scale have accompanied continual environmental changes.... The cause of these mass extinctions is still very much in doubt and constitutes a major problem of evolutionary history."[24]

Eric Larrabee, discussing the recent revival of interest in catastrophism and, in particular, the type of catastrophism proposed a decade ago by Immanuel Velikovsky, appropriately observes:

> "The nineteenth century found it natural to think in terms of continuity and reassurance, of slow evolution and gradual processes undisturbed by sudden and unpredictable disruptions. We of the twentieth have known a different universe, have seen the overturning of stability in every sphere, have come to live from day to day with the constant threat of violence unimaginable. For us catastrophes are less difficult to visualize...."[25]

Now since catastrophes of tremendous severity must certainly be invoked to explain most of the geological deposits and formations, the next question is how *many* such catastrophes are involved. Application of Occam's Razor would suggest that the smallest possible number of such catastrophes that can explain the data would provide the best hypothesis. If it should be at all conceivable that only *one* great catastrophe, with many more or less simultaneous concomitant effects, could suffice for the purpose, then this should very seriously be considered as the most reasonable of all possible explanations for the fossil record. And this, of course, brings us to the third great historical fact revealed by the Bible, namely the universal deluge of the days of Noah.

---

[24] "Crises in the History of Life," *Scientific American,* Vol. 208, February, 1963, p. 77.

[25] "Scientists in Collision: Was Velikovsky Right?", *Harper's Magazine,* August, 1963, p. 55.

Chapter IV

Water and the Word

Strikingly significant is the fact that nearly all of the earth's sedimentary rocks and their contained fossils have been deposited under moving water, in stratified beds, and that there are many indications that this deposition was catastrophic. It is clearly more than coincidental that the Biblical revelation describes a great aqueous catastrophe in which "the world that then was, being overflowed with water, perished." Furthermore, the divinely inspired record of the flood in the Bible is corroborated by hundreds of reflections of this same great event handed down in the legends and historical records of practically all nations and tribes in the earth. It is eminently scientific and reasonable, therefore, to consider with full seriousness the proposition that the fossil record may be in large measure a record of the effects of the Noahic flood.

This was, in fact, exactly the interpretation placed on the fossil record for several generations prior to the time of Lyell and Darwin. The philosophy of uniformity and evolution later displaced "flood geology," but this was more on the basis of philosophical preference than scientific necessity. In view of all the considerations outlined above, especially the scientific weakness of the theory of evolution and of geological uniformitarianism, we propose that a return to a clear-cut doctrine of Biblical creationism and catastrophism is amply justified at this time. It is recognized that there are many serious problems involved in such a position, and that we do not by any means have all the answers to these problems as yet, but the basic framework seems much more realistic than that of evolutionary uniformitarianism, and the problems and objections much less serious.

Space will only allow a brief discussion of the flood and its effects. For this discussion, of course, the Bible will be taken as basic and as absolutely and literally true. This is admittedly a pre-supposition, from which we shall then proceed to deduce certain conclusions; the latter will then be compared with the known geological phenomena. This is exactly parallel to the procedure of the evolutionist, whether or not he will acknowledge it, since he starts from the pre-supposition that evolution is true and then proceeds to interpret the phenomena in accordance with that assumption.

Our discussion can be expeditiously pursued in terms of a Biblical cosmology centered around the theme of water, which is the most important substance in the material earth and which, of course, was the primary agent in the flood and in the deposition of the fossils and the sedimentary rocks.

The third chapter of the second epistle of Peter provides a most remarkable outline for this study, and we shall therefore quote verse 3 through 13 of this chapter below:

> "Knowing this first, that there shall come in the last days scoffers, walking after their own lusts, (4) And saying, Where is the promise of his coming? for since the fathers fell asleep, all things continue as they were from the beginning of the creation. (5) For this they willingly are ignorant of, that by the word of God the heavens were of old, and the earth standing out of the water and in the water; (6) Whereby the world that then was, being overflowed with water, perished: (7) But the heavens and the earth, which are now, by the same word are kept in store, reserved unto fire against the day of judgment and perdition of ungodly men. (8) But, beloved, be not ignorant of this one thing, that one day is with the Lord as a thousand years, and a thousand years as one day. (9) The Lord is not slack concerning his promise, as some men count slackness; but is long-suffering to usward, not willing that any should perish, but that all should come to repentance. (10) But the day of the Lord will come as a thief in the night; in the which the heavens shall pass away with a great noise, and the elements shall melt with fervent heat, the earth also and the works that are therein shall be burned up. (11) Seeing then that all these things shall be

> dissolved, what manner of persons ought ye to be in all holy conversation and godliness, (12) Looking for and hasting unto the coming of the day of God, wherein the heavens being on fire shall be dissolved, and the elements shall melt with fervent heat? (13) Nevertheless we, according to his promise, look for new heavens and a new earth, wherein dwelleth righteousness."

Notice that this passage divides cosmology into three distinct historic periods: (1) the heavens and the earth which were "of old," in verse 5; (2) the heavens and the earth which "are now," in verse 7; and (3) the heavens and the earth which we are still looking for and which shall be "new," as mentioned in verse 13. These three cosmologies are separated by two great world-wide catastrophic events: (a) the great flood, which destroyed the old heavens and earth, and (b) a judgment by fire, perhaps atomic disintegrations, at the day of the Lord, which shall destroy the present heavens and earth. Let us now consider briefly each of these three cosmologies, and especially the role played by water in each.

The creation of the original earth, with its atmospheric heavens, as described in verse 5, obviously carries us back to the creation recorded in Genesis 1. This was accomplished "by the word of God," not by processes of evolution. It has already been pointed out that present processes can give us no information concerning the processes used by God in creation; any such information must necessarily come by divine revelation, and not by application of the uniformity principle. In this verse is contained just such a revelation. The earth was formed "standing out of the water and in the water." The American Standard Version translates this phrase: "compacted out of water and amidst water." Numerous other translations have been suggested and it is likely that the reason it has seemed to be such a difficult phrase to translate is that we subconsciously try to impress our experience of present physical phenomena back upon these events which took place in the creation period, and this is not legitimate. In some way, the initially formless earth was caused by the divine Word to be "compacted out of the water." This most probably describes the primeval condition when the Spirit of God moved over the face of the waters comprising the "deep"

(Genesis 1:2) and caused a "division of the waters from the waters." First God caused a division of the "waters under the firmament" from the "waters above the firmament" (Genesis 1:6, 7) and then He caused a division of the waters under the firmament from the "dry land" (Genesis 1:9, 10). The "firmament" was merely the atmosphere, called "heaven," in which birds were later to fly, and light from the sun and moon to be scattered and diffused (Genesis 1:8, 15, 20).

In order for the upper waters to be maintained aloft by the gases of the lower atmosphere and also to be transparent to the light of the sun, they must have been in the form of a vast blanket of water vàpor, extending far out into space, invisible and yet exerting a profound influence on terrestrial climates and living conditions. Such a "canopy" would have caused a world-wide warm, mild climate, with only minor seasonal and latitudinal differences. This in turn would have inhibited the great air circulational patterns which characterize the present atmosphere and which constitute the basic cause of our winds, rains, and storms.

There could have been no rain in the form with which we are familiar, and this is exactly the testimony of Scripture (Genesis 2:5, 6). But there was a system of rivers and seas (Genesis 1:10; 2:10-14), nourished probably by waters that had been confined under pressure beneath the lands when the lands and the waters were separated as well as by the low-lying vapors that were daily evaporated and recondensed (Genesis 2:6). These rivers, especially one which emerged from a great artesian spring in the Garden of Eden (Genesis 2:10), were the main sources of water for Adam and his descendants.

The vapor canopy also would have served as a highly effective shield against the many powerful and harmful radiations that surround the earth, and which are now only partially filtered by our present atmosphere. Such radiations are now known to be the cause of much damage to man's genetic system, tending to cause harmful mutations and general biological deterioration. It is quite possible that the healthful environment created by this great thermal vapor blanket was one major factor contributing to human longevity in those days.

But this pristine environment was disturbed by corrupting influences. "Sin entered the world, and death by sin" (Romans 5:12). The corrupting leaven spread throughout the world until almost "the whole was leavened." The Bible says that "God looked upon the earth, and, behold, it was corrupt; for all flesh had corrupted his way upon the earth" (Genesis 6:12). Therefore, God finally said that He would "bring a flood of waters upon the earth, to destroy all flesh, wherein is the breath of life, from under heaven; and everything that is in the earth shall die" (Genesis 6:17).

The very waters out of which the earth had been compacted and which had sustained its life and pleasant climatic conditions were now to cause its destruction. "Whereby," by this same water, as Peter says, "the world that then was, being overflowed with water, perished" (II Peter 3:6). The great expanse of waters above the firmament was condensed and plunged to the earth, continuing everywhere at fullest intensity for forty days and forty nights (Genesis 7:12). The "great deep," including the waters of the seas and also that part of the primeval deep which had been locked as vast storehouses of waters beneath the rocks of the earth's crust, also issued forth, as "all the fountains of the great deep were broken up" (Genesis 7:11). This latter upheaval must have been accompanied by the eruption of subterranean magmas, and these by great earthquakes, and these in turn by tremendous tsunami waves in the seas. Destruction beyond imagination must have been wrought on the antediluvian earth! The waters rose until "all the high hills, that were under the whole heaven, were covered," and "the mountains were covered," and "all flesh died that moved upon the earth, both of fowl, and of cattle, and of beast, and of every creeping thing that creepeth upon the earth, and every man" (Genesis 7:19-21). Once again, as in the beginning, there was a universal ocean. The same waters that had sustained the life of the world now became its shroud.[1]

---

[1] This discussion has been taken in part from an article "Water and the Word," by the present author, in *Bibliotheca Sacra,* Vol. 118, July, 1961, pp. 203-215.

In view of such statements by the Biblical writers, it is a remarkable testimony to man's perverse reluctance to accept the Word of God in preference to the reasonings of man that Christian expositors could ever have proposed the "local-flood-theory." The Genesis account so clearly records the fact of a world-wide, uniquely destructive deluge that it would hardly be possible to present such a concept in any clearer and more emphatic terms than are actually recorded. There are at least thirty distinct references in Genesis 6 through 9 to the flood's universality! All the mountains under the whole heaven were covered with water, for a year; everything in the dry land died; mankind was to be destroyed *"with the earth"* (Genesis 6:13; 9:11); and the ark was to be constructed as a special vehicle for the very purpose of keeping seed alive upon the face of the earth. These prescriptions cannot possibly be understood except in terms of a universal flood. The very fact of the ark is itself sufficient to demonstrate this, for it would have been absurdly unnecessary for refuge from a local flood. In the one hundred twenty years of building the ark, Noah and his family, not to mention the birds and the beasts, could have migrated far beyond the utmost limits of any conceivable local flood.

With the precipitation of the vapor canopy, there was no longer the world-wide mild climate that prevented the development of winds and storms. Soon great winds began to blow (Genesis 8:1), generating great waves and currents (Genesis 8:3); perhaps these forces also triggered the tectonic forces which must have been acting when "the waters hasted away [the mountains rose, the valleys sank down] unto the place which thou hadst founded for them. Thou hast set a bound that they may not pass over; that they turn not again to cover the earth" (Psalm 104:6-9, ASV).

An entirely different climatic mechanism henceforth prevailed. Distinct seasons were inaugurated (Genesis 8:22), and the rainbow was established (Genesis 9:13), neither of which was possible with the antediluvian vapor canopy. Furthermore, human life spans began to decline, probably as a result of the increase in atmospheric radiations and the general austerity of climate and living conditions. The "heavens and the earth which are now" involve an entirely new, though still quite effective, hydrologic cycle.

The oceans are now much larger than they were before the flood, since they now contain the waters that were formerly "above the firmament," as well as those released through the "fountains of the great deep." It is these that now constitute the great "storehouses" of water that are essential for the operation of the present hydrologic cycle (Psalm 33:7). Waters are evaporated from the oceans (Psalm 135:7), carried inland by the winds (Ecclesiastes 1:6), caused to encounter particles of dust and sea salt to serve as nuclei of condensation (Proverbs 8:26), condense into liquid water droplets in the form of clouds (Job 26:8), which in turn under the proper conditions coalesce and fall as rain (Job 36:27-28), providing water for maintenance of life on the earth (Isaiah 55:10), and finally return by the rivers to the oceans from which they come (Ecclesiastes 1:7).

But this present cosmology is being "kept in store" (II Peter 3:7), perhaps an implicit reference to the principle of mass and energy conservation, for another great world-wide destruction in the "day of the Lord," when both the present earth and its atmospheric heavens will be melted with "fervent heat." This judgment, like that of the flood, will be universal and will result in a completely changed environment once again. And this time, there will be a "new heavens and a new earth, in which dwelleth righteousness" (II Peter 3:13).

The present great water storehouse will no longer be needed, and there shall be "no more sea" (Revelation 21:1). As the present great and wide sea (Psalm 104:25) should have been to mankind an ever-present reminder of God's former judgment by water, so the Lake of Fire will be an eternal reminder of God's greater and final judgment by fire.

There is no need or justification for "spiritualizing" the promises concerning the new heavens and the new earth, any more than the account of the first creation should be spiritualized, or the Biblical descriptions of the present cosmology need to be spiritualized. In eternity, there will not only be new *heavens* but also a new *earth,* and this can hardly refer to anything other than this planet on which man now lives. However, the earth will be "made new" (that is, *renovated*) and will be perfectly suited as an abode for righteousness. The curse will be removed (Revelation 22:3), so there will be no evidence of the effects of sin, disorder, decay, and death. The second law of thermodynamics will no longer control physical processes. Possibly there may be a restoration of the original vapor canopy around the earth. As far as the Scriptures indicate, the earth's needs for water will be fully satisfied by the "pure river of water of life" (Revelation 22:1), which proceeds from the throne of God and of the Lamb.

Regardless of how precisely we are able to interpret these properties of the future earth and atmosphere, however, the important point that needs to be stressed here is that the Bible emphatically teaches, in both the Old and New Testaments, that the antediluvian cosmology was completely and globally changed by the great deluge. The tremendous torrents of rain, the volcanic upheavals, the complex of destructive tsunamis, the raging streams of water, the tectonic movements – all must have exerted such a profound influence on the earth's prior economy that, as Peter said, "the world that then was, being overflowed with water, perished."

The flood, then, must mark a great discontinuity in the ordinary geological and hydrological processes of the earth. Any geological deposits which may have existed before the flood must have been eroded, re-worked and re-deposited in some new location and sequence, such that it would not be possible to deduce now, by analogy with present processes, the geologic history of the antediluvian earth. Pre-flood geography must have been completely revised, with new continents formed by great uplifts after the flood, as well as new ocean basins for the retention of the mass of flood waters covering the earth. It follows as plainly as anything could be that any system of historical geology which ignores the fact and effects of the universal flood must be largely fallacious. Geologic dating methods which assume continuity of rates of the associated geologic processes, as all such methods necessarily do, can only be valid for a period of time extending, at most, back to the flood.

All of this does not mean, however, that the flood deposits

would be a completely heterogenous and confused mass, with no order and no possible system for interpretation. Although the destruction and upheavals were world-wide, the basic physical processes (as controlled by the laws of thermodynamics, by gravity, stress-strain relations, the laws of motion, etc.) were of course not affected. Out of all the apparent chaos of destruction that must have been caused by the flood, it is apparent that the final deposits which resulted from it would have assumed a certain statistical order.

For example, there would naturally be a tendency for those sediments and organisms which occupied the lowest elevations before the flood to be buried deepest by the flood. Thus, simple marine organisms and marine sediments would tend to be buried deepest, then fishes and more complex marine creatures, then reptiles and amphibians, then mammals, and finally, man.

Another factor controlling to some extent the order of deposition of the sediments and the organisms contained in them would be that of the relation between the specific gravity and the hydrodynamic drag. Each particle of material, as well as the remains of each animal, would tend to fall by gravity out of the aqueous mixture in which it was being carried. This tendency would be resisted by the hydrodynamic drag force of the water acting upward on it. The latter depends on the state of turbulence of the water and also on the shape of the object, being greatest for objects of complex shape and least for objects of streamlined shape. Thus, there would be a tendency for organisms of high density and simple structure to settle out most rapidly and, therefore, to be buried deepest. This factor of hydrodynamic selectivity is often highly efficient and would tend to cause rather highly sorted sediments and fossils, with organisms of similar size and shape being buried together.

A third factor which would have an important effect, so far as living organisms were concerned, would be their relative abilities to escape the onrushing flood waters by retreating to higher ground. The simpler, less mobile, smaller creatures would thus be caught and trapped first, whereas higher animals, and especially man, would often be able to retreat to the very highest points in the region before being inundated. This, too, would mean that most men and higher animals would never be buried at all in sediments, but would float on the surface of the waters until decomposed or destroyed by scavenger fish.

Many other criteria might be deduced for harmonizing the data from the resulting flood deposits, but it is significant that these three important factors would all tend to produce a fairly well graded series of deposits, increasing in size and complexity with increase in elevation of burial. And, of course, this is exactly the general order found in the fossil record! There are many exceptions, as would be expected from the complex of phenomena associated with the great deluge, but this is most definitely the statistical order. Thus, the fossil deposits and the sediments in which they are contained can very logically be viewed as an actual historical record, preserved in tablets of stone, of the terrible events of the year of the great flood.

We conclude this section, then, by noting that the great flood offers a plausible solution to the enigma of the fossil record. Although there are undoubtedly many difficulties[2] that may appear in attempting to understand particular geological formations in terms of the flood, these difficulties are not nearly so great as in attempting to interpret them consistently in terms of uniformity and evolution. If one is willing simply to reject his innate bias in favor of uniformitarian explanations and to accept the Biblical framework by an act of faith, centered around the three great facts of revealed Biblical history – namely, a real creation of a fully functioning and harmonious universe, the fall of man and resulting curse of God on the whole creation, and the great universal deluge in the days of Noah – then he will find that the entire scope of scientific data will begin to become meaningful and inter-related instead of confused and contradictory. The universe will become a living testimony to the presence of a God of both love and judgment. "In Christ are hid all the treasures of wisdom and knowledge" (Colossians 2:3).

---

[2] The most serious of these difficulties are dealt with in considerable detail in *The Genesis Flood* (by John C. Whitcomb and Henry M. Morris, 331 ff.), and shown to be fully capable of resolution and harmony with Biblical creationism and catastrophism.

Chapter V

The Origin of Evolution

The idea of evolution did not, of course, originate with Darwin. Its modern widespread acceptance does date from the publication of Darwin's *Origin of Species,* but it was a doctrine held by many scientists and philosophers before Darwin. Belief in spontaneous generation of life from the non-living and in transformations of the species was quite common among the ancients. Among the early Greeks, for instance, Anaximander taught that men had evolved from fish and Empedocles that animals had been derived from plants.

The doctrine of spontaneous generation was well-nigh universal among the ancients. It was commonly thought that not only insects and fishes, but probably also the higher animals and man himself, were on occasions generated directly from mud or slime or some other inorganic medium. And if such great marvels as this could and did occur, there was no great problem in believing that one species could be transmuted into another.

As Paul Amos Moody remarks:

> "Ideas that by one means or another evolution does occur far antedated Darwin, however. In fact, such ideas are probably as old as human thought."[1]

One striking fact emerges from the study of all the ancient cosmogonic myths, whether from Babylon, Greece, Egypt, India,

[1] "Introduction to Evolution" (2nd Ed. New York, Harper & Brothers, 1962), p. 3.

or wherever. The concept that the universe had originally been *created,* out of nothing, by an act of God, is completely absent. Always there is a primeval chaos or a primeval system of some kind, upon which the "gods," or the forces of nature, begin to work in order to bring the world and its inhabitants into their present state. Special creation seems to have been a doctrine completely unknown (or, if known, rejected) by the ancients.

Often the naive assertion is made by modern skeptics that the Genesis account of creation was written as an "accommodation" to the simple culture of the early Hebrews — that it could not have been presented in the framework of evolution because they were incapable of comprehending an evolutionary system of origins. Such a notion as this is ridiculous on its face, for the simple reason that all early peoples were accustomed to think *only* in evolutionary terms. The idea of special creation — that is, creation *ex nihilo* by an omnipotent, eternal God who alone existed "in the beginning" — was a tremendously new and revolutionary thought! Further, the idea of a *recent* creation of the universe, measurable in terms of genealogical records kept since the first man, would have been much harder for the ancients to accept than the idea of billions of years of geologic time would have been. It is well known that many ancient astronomers and philosophers dated the universe almost infinitely old. Thus, the Biblical revelation of origins was unique in the ancient world, and was in fact almost universally resisted and rejected until the supposed triumph of Christianity in Europe. And even this apparent triumph was short-lived, as evolutionary speculations continued to thrive even within the church, preparing the way finally for the emergence of modern Darwinism in the 19th century.

As far as the modern world is concerned, we have already seen at some length that the philosophy of evolution completely dominates modern thought, whether in the natural sciences, the social sciences, philosophy or even religion. It is considered so basic that, as Huxley says, "the whole of reality *is* evolution." The Biblical concept of special creation of all things only a few thousand years ago is widely derided as so ridiculous that only the hopelessly ignorant or prejudiced could possibly believe it.

Yet, despite the almost universal prevalence of evolutionary thinking, both in the past and in the present, it is completely contrary to all true science as well as to Biblical revelation, as we have shown at some length in the foregoing pages. This is surely a most amazing state of affairs!

If evolution is basically impossible from a scientific point of view (as demonstrated by the universality of the two laws of thermodynamics) and untrue from a historical point of view (as demonstrated by God's revelation of a finished creation and subsequent curse on creation), then how can we explain the well-nigh universal insistence that all things must have come about by evolution? The real answer, we suggest, is found in II Corinthians 4:3, 4:

> "But if our gospel be hid, it is hid to them that are lost: In whom the god of this world hath blinded the minds of them which believe not, lest the light of the glorious gospel of Christ, who is the image of God, should shine unto them."

The answer is *Satan!* He has blinded the minds of men with respect to the gospel. The gospel is the good news of a Saviour, who has borne the sins of men on the cross, that all who believe on him might be saved. But if men have evolved by natural processes out of the elementary "stuff" of the universe, then there is no responsibility to a Creator, there has been no fall and no curse and, therefore, there is no need of a Saviour!

The "great dragon . . . that old serpent, called the Devil, and Satan," who "deceiveth the whole world" (Revelation 12:9) — must without any doubt be the one who has fathered this monstrous lie of evolution, for he is the father of lies. The Lord Jesus called him the "prince of this world" (John 12:31; 16:11); the Apostle Paul called him "the prince of the power of the air, the spirit that now worketh in the children of disobedience" (Ephesians 2:2); the Apostle John stressed that "the whole world lieth in the wicked one" (I John 5:19), and Satan himself was not contradicted by Jesus when he told Him "All this power [i.e., all the kingdoms of the world] is delivered unto me" (Luke 4:6).

When one recognizes the Satanic origin of evolution, then many otherwise confusing issues begin to come into focus. The ultimate issue in the universe, in fact the only real issue, is that of God's sovereignty versus the asserted autonomy of his creatures. Is God really the Creator and King of the universe, or is he limited to greater or less extent by his creatures?

There are not *many* religions and philosophies among men. There is really only one, and that is the rebellious and blasphemous belief that autonomous man is capable of controlling his own destiny independently of the will of his Creator. Every religion (other than Christianity) is an attempt on man's part to earn "salvation" or to improve his standing in the world, either temporally or eternally. Every non-Christian philosophy is an attempt to deduce ultimate truth concerning the universe without submission to the revealed Word of God. All of man's religions and philosophies, apart from the grace of God revealed in his Word, are *man-centered* – or, perhaps more generally, *creature-centered* – rather than *Creator-centered.* They all involve some system of works, of improvement, of development, of human betterment, of *evolution!* – rather than simple submission in helpless faith to the sovereign grace of God manifest in the sacrifice of the Lamb of God for the sins of the world.

It is perhaps providential that, in the generally adopted system of verse divisions, the middle verse of the Bible is the simple statement of Psalm 118:8 that:

> "It is better to trust in the Lord than to put confidence in man."

In a very real sense, this is really *the* message of God to man. It is better to trust in God to provide salvation than in any system of human works. It is better to believe the revealed Word of God than any science or philosophy devised by man.

And since man and all of man's works are now under the dominion of the god of this world, Satan, putting confidence in man means, in the last analysis, putting confidence in Satan. Doubting God's Word is tantamount to accepting Satan's word. As Paul says "But I fear, lest by any means, as the serpent beguiled Eve through his subtility, so your minds should be corrupted from the simplicity that is in Christ" (II Corinthians 11:3).

But Satan, with all his power, is like man, only a creature! He is not omnipotent, nor omniscient, nor omnipresent, as is God.

How is it, then, that he is able to deceive the whole world, and even a host of angels who follow him in rebellion against the Creator, God? How can any number of mere creatures, even when led by such a powerful member of creation as Satan, ever possibly hope to be ultimately successful in a revolt against God? This seems, on its very face, to be the height of insanity!

According to the Bible,[2] Satan originally was "full of wisdom and perfect in beauty." He was "the anointed cherub that covereth – perfect in [his] ways" (Ezekiel 28:12-15). But "thine heart was lifted up because of thy beauty, thou hast corrupted thy wisdom by reason of thy brightness" (Ezekiel 28:17). Satan said in his heart: "I will ascend into heaven, I will exalt my throne above the stars of God: ... I will be like the most High" (Isaiah 14:13-14).

Because of this rebellion, in which he was accompanied by a great host of other created beings (perhaps "a third part of the stars of heaven" as intimated in Revelation 12:4), he was "cast to the ground" (or "earth" – Ezekiel 28:17), and will ultimately be "brought down to hell" (Isaiah 14:15).

But *why* did this one, the Anointed Cherub, Lucifer, choose to become the Adversary, Satan? The answer, of course, is pride. His "heart was lifted up," and he desired to usurp the place of God in the universe.

But, again, *why* did he allow this pride to deceive him into thinking that such a desire could possibly be fulfilled? He knew that his great beauty and wisdom were his only because God had, in grace, given them to him. He knew that he was only a creature of the eternal God, in spite of all his exalted position.

---

[2] Commentators are in disagreement as to whether these passages (Ezekiel 28:12-19 and Isaiah 14:2-15) apply specifically to Satan or only to his human instruments, the King of Tyre and King of Babylon respectively. The present writer believes that they apply initially to the earthly kings, but ultimately and more fully to the powerful spirit of evil energizing these men, and therefore to Satan himself, since many of the statements made seem impossible of specific application to any human being. In either case, the spirit of rebellion and pride described in these passages is surely Satanic in nature and graphically illustrates the character of his revolt against his Creator.

Or did he? To be sure, God had told him these things. He was not in existence from eternity, as was God, for there had been *a day when he "was created"* (Ezekiel 28:13, 15). But there must have come a time when he began to wonder if this were really so. After all, the only evidence he had that this was so was God's Word. Perhaps, in order to keep him satisfied in a secondary role, God had merely *told* him that He had created him. Could it not be that both he and God had come into existence in some way unknown and that it was just an accident of priority in time that enabled God to exercise control? With all his beauty and wisdom, he could undoubtedly win the allegiance of many other angels who had similar reason to question the Word of God.

And so pride begets unbelief and unbelief strengthens pride and this twin sin of unbelieving pride becomes the root sin of all other sins. It led to Satan's rebellion, and later to Adam's fall, and is the very foundation of all the world's sin and suffering from that day to the last day.

How could Satan possibly rationalize this notion that both he and God and all other beings had come into existence in similar fashion and therefore were of essentially the same order? If God had not created him, who had? If God were not all-powerful, who was? In other words, who was *really* God? The only possible answer that could be given by Satan which would in any way rationalize his rebellion was that there was really no Creator at all! Somehow everything must have come about by a process of natural growth, of development, of *evolution.* If he would *not* believe the Word of God, then this is what he *must* believe.

And this is what he *still* believes! For, despite the clear testimony of the Word of God concerning his ultimate defeat and eternal punishment (and Satan thoroughly knows the Scriptures), he still refuses to believe that it is really so, and so continues to rebel and hope that he will ultimately be victorious in this conflict of the ages.

How utterly, fantastically, hopelessly, tragically foolish! "The *fool* hath said in his heart, there is no God" (Psalms 14:1; 53:1). And this same tragic foolishness was later reproduced by

Satan in Adam and in Adam's children. Because, after all, any doubting of the Word of the Creator—or any insertion of creaturely reasoning or efforts in lieu of the revealed grace of God—is, in the last analysis, a denial that God is *really* God.

This must be the original source of the theory of evolution. This is the only possible explanation of origins apart from God. Some kind of process of evolution will, therefore, be found at the core of every religion or philosophy of man, apart from the Word of God, as revealed in the Holy Scriptures and in the Living Word, Jesus Christ.

The sin of Adam and Eve, essentially, was believing the serpent's suggestion that God's Word was not to be trusted, that he was withholding knowledge which they had a right to have, and that they also could be "as gods, knowing good and evil" (Genesis 2:5). This was the same old deception with which Satan had deceived himself. Their desire for "knowledge" (one might say "science") led to their fall and the resultant curse of God on man and on his "dominion," which had now passed under the apparent control of Satan.

Since the curse, the history of mankind has been a history of almost continual deterioration, interrupted only by occasional infusions of regenerative power by God, in grace, upon those men whom he chose. The tragic story is summarized in Romans 1:21-25:

> "Because that when they knew God they glorified him not as God, neither were thankful; but became vain in their imaginations, and their foolish heart was darkened. Professing themselves to be wise, they became fools, And changed the glory of the uncorruptible God into an image made like to corruptible man, and to birds, and fourfooted beasts, and creeping things. Wherefore God also gave them up to uncleanness through the lusts of their own hearts, to dishonor their own bodies between themselves: Who changed the truth of God into a lie, and worshipped and served the creature more than the Creator, who is blessed forever. Amen."

This searing passage describes the all-but-universal drift of the pagan world of antiquity from a primal acknowledgment of the *Creator* into a degenerate system of evolutionary pantheistic humanism. This is the basic framework of all man-centered, as opposed to Creator-centered, religion. It is evolutionary because it rejects creation *ex nihilo* and always accounts for the world as somehow developing from pre-existent materials; it is pantheistic because it identifies God with nature in one way or another; and it is humanistic because it exalts man's reason above the revealed Word of God.

And, now, what is the difference between this ancient paganism and our modern evolutionary philosophy? The answer is that there is no essential difference whatever! It is the same lie of Satan, perhaps in a more sophisticated garb, to suit modern tastes, but there is really no difference. Modern scientism is evolutionary, because it seeks to account for all things in the universe in terms of natural developmental processes from pre-existent materials. It is pantheistic because God is identified with and limited by his "creation." And it is humanistic because it "worships and serves the creature more than the Creator." Man, as the highest stage to which evolution has yet attained, is in essence his own God.

This is the state of things in the intellectual world today, and this is the situation foretold by the Apostle Peter as prevalent in "the last days." He said that the latter-day scoffers would jeer: "Where is the promise of his coming? For since the fathers fell asleep, all things continue as they were from the beginning of the creation" (II Peter 3:3, 4).

That is, men in the last days will resist the very thought that the Creator might one day come in judgment, requiring them to give an account of their stewardship. In order to rationalize away such a notion, they would propound a doctrine of "uniformitarianism," insisting that "all things continue as they were," and that, therefore, there is no cause to be worried over a possible future suspension of those processes. In support of this contention, they assert that these processes have continued unchanged since the very "*beginning* of creation." This can only mean that creation itself was accomplished by these same processes, and not by any divine creation *ex nihilo*. In effect, this is again nothing but the same old denial that God is Creator and, therefore, that God is God. Therefore, he cannot "come" again as he had promised, in judgment.

In refutation of this false and wicked philosophy, the apostle merely reminded the believers of the two great revealed facts of a primeval creation by the Word of God and of a subsequent world-destroying judgment when "the world that then was, being overflowed with water, perished." Both of these facts completely and irrefutably discredit the doctrine of uniformitarianism and evolutionism advocated by the last-day scoffers.

In summary of this section, we have seen that the origin of all the evil in the universe must have been coincident with the origin of the idea of evolution, both stemming from Satan's rejection of God's revelation of himself as Creator and Ruler of the universe. This primal act of unbelief and pride led to Satan's fall; the same basic act of unbelief and pride later led to the fall of man. Similarly, unbelief in God's Word and man's pride in his own ability to rule his own destiny have yielded the bitter fruits of these thousands of years of human sin and suffering on the earth. And today, this God-rejecting, man-exalting philosophy of evolution spills its evil progeny – materialism, modernism, humanism, socialism, Fascism, communism, and ultimately Satanism – in terrifying profusion all over the world.

Chapter VI

The Death of Evolution

In view of the world-wide prevalence of the evolutionary philosophy in modern times, not to mention its age-long priority in the thought-structure of men rebelling against their Creator, it may still seem strange to the reader that one would suggest such a title as "The Twilight of Evolution" for this book. Especially is this so since much of the book has been devoted to pointing out what seems to be the very opposite, namely, that evolution is now almost universally accepted as the basic framework for almost all disciplines of human study and research. A more fitting title, it might seem, would be something like "The Triumph of Evolution."

But there are two definite ways in which the selected title is singularly appropriate. In the first place, the whole area of evolutionary "science" seems to lie in a strange twilight zone. Its whole premise contradicts not only divine revelation but also basic scientific law (e.g., the two laws of thermodynamics). Its numerous suppositions and speculations evaporate away whenever the warm light of either Biblical or scientific *facts* is brought to bear on them. Its seemingly imposing arrays of evidence turn out to be nothing but shadows when closely examined.

But evolution is also in its twilight period chronologically speaking. This may not yet be obvious, but nevertheless its night is fast approaching.

In spite of the tremendous pressure that exists in the scientific world on the side of evolutionary propaganda, there are increasing signs of discontent and skepticism. As noted before, all the important scientific periodicals and all the major scientific book publishing houses are completely under the dominance of evolution. Somehow the fiction that evolution is a proven fact and that no one but a scientific obscurantist could possibly question it has become so powerful a club that these outlets will hardly touch anything remotely savoring of creationism. Furthermore, it would be difficult or impossible for a known anti-evolutionist to get or to retain a position on at least a majority of the major university faculties. And it is extremely difficult for a known anti-evolutionist or anti-uniformitarian to obtain a doctoral degree in such fields as geology, biology, anthropology, and sociology.

But in spite of all these and other pressures, it is certain that there are great numbers of well-educated and well-informed people in many areas of science today who do *not* believe in evolution and who *do* believe in the Biblical accounts of creation and the flood. These may not often be willing or able to make their views widely known, but they do exist, nevertheless. In his own relatively small circle of acquaintances, the writer personally knows men with the Ph.D. degree in geology, in biology, in anthropology, physics, chemistry, astronomy, entomology, hydrology, mathematics, genetics, archaeology, and other sciences – as well, of course, as numerous men with Ph.D.'s in engineering – who do not believe in evolution. There is obviously nothing in any of these sciences, therefore, that really *compels* their practitioners to accept the supposed "fact" of evolution. The latter is nothing but a propaganda device which has been used – and very effectively – to awe people into accepting it.

Here and there, surprisingly enough, even in the standard scientific publications media, there are beginning to appear evidences of doubts concerning evolution. Nothing much which is overtly skeptical of evolution as a whole can be published, of course, but at least signs are appearing which indicate that there may exist a very substantial substratum of doubt concerning evolution today.

For example, one of the speakers at the Darwinian Centennial Celebration, Everett C. Olson, who is Professor of Geology at the University of Chicago, made the following comment:

> "There exists, as well, a generally silent group of students engaged in biological pursuits who tend to disagree with much of the current thought but say and write little because they are not particularly interested, do not see that controversy over evolution is of any particular importance, or are so strongly in disagreement that it seems futile to undertake the monumental task of controverting the immense body of information and theory that exists in the formulation of modern thinking. It is, of course, difficult to judge the size and composition of this silent segment, but there is no doubt that the numbers are not inconsiderable."[1]

Dr. Olson was not saying that this group of scientists were anti-evolutionists, of course, but only that they rejected the dominant theory of evolutionary mechanisms held by such men as Huxley, Simpson, Mayr, *et al.* This dominating theory, commonly known as the "modern evolutionary synthesis," involves the factors of genetic mutation, Mendelian inheritance, population drifts, and the critical role of natural selection. This is the more-or-less standard theory of evolution as promulgated by the large majority of evolutionary spokesmen today. It completely rejects any form of creative or causative principle in evolution, as well as any effect of "saltations," or "jumps," in evolutionary progression. The great strength of this theory is that it recognizes that there is no basis in the biology of present processes for any kind of biologic change except very small mutations, Mendelian variation and natural selection.

But many biologists feel that this concept tends to ignore the significance of the very distinctive and discrete character of the different kinds of organisms, both past and present. The "synthetic" theory tends to regard all units, whether species, genera, or others, as only temporary and fluctuating, so that any attempt at classifying these organisms must be cast in an evolutionary mold. The school of thought of which Olson speaks, on the other hand, feels that classification is a science which is essen-

---

[1] Morphology, Paleontology, and Evolution" in *The Evolution of Life* (Sol Tax, Ed., University of Chicago Press, 1960), p. 523.

tially independent of any particular theory of evolution and, in fact, independent even of the supposed fact of evolution.

From the vantage-point of the anti-evolutionist, however, this controversy is obviously occasioned by the fact that there is no real evidence for evolution in present genetic processes. These processes simply do not account for the present kinds of organisms, with their clear-cut "gaps" between species, genera, families, etc. Consequently, this school of "systematic classificationists" or "typologists" prefers to do its work quite independently of the embarrasing restrictions imposed upon it by some specific theory of evolution. Thus, although neither Olson nor most of the silent segment of whom he speaks would be bold enough to question the *fact* of evolution, yet, the very fact of their recognition that the evolutionary synthesis cannot really account for the known facts of biology with which they deal, is real grounds for an encouragement to think that many may be on the verge of recognizing the fallacy in the very idea of evolution itself.

Two systematic biologists at Stanford University have recently been able to publish a provocative paper along this line. They stress the circular reasoning involved in the popular approach to evolutionary synthesis as follows:

> "For nearly 200 years taxonomists have followed Linnaeus in arranging organisms in 'natural groups.' Darwin supplied a rationale for the existence of such groups, and in the minds of many workers the existence of groups and their probable cause have become inseparable. This has led to the so-called phylogenetic approach to taxonomy, in which, in the absence of satisfactory fossil records, taxonomic systems often are used as the basis for constructing phylogenetic trees. Unfortunately, these trees sometimes are then employed to alter the original taxonomic system. This circular procedure produces systems with some predictive value and information content, although the process of creating these systems through repeated revision is time-consuming and relatively inefficient."[2]

It may also be noted in passing that this method of building phylogenies (that is, supposed evolutionary histories of particu-

[2] Paul R. Ehrlich and Richard W. Holm: "Patterns and Populations," *Science,* Vol. 137, August 31, 1962, p. 655.

lar organisms) is so constructed that it is practically impossible to refute it. This is because of the circular reasoning implicit in the development of such a phylogeny. The phylogeny is built up on the basis of what the particular evolutionist thinks it ought to be, and then this phylogeny is taken as proof that the organism has evolved in that manner, and that this proves that the method employed is correct! Olson says:

> "That no organic event has been discovered that cannot be explained by the 'synthetic theory,' or selection theory, as is often stated, is in a sense true. On the other hand, the feeling of a slight sense of frustration in the elasticity involved in developing a universal explanation is hard to avoid, a feeling somewhat in sympathy with V. Bertalanffy (1952) when he noted 'a lover of paradox could say that the main objection to selection theory is that it can't be disproved.' . . . In this sense, there is little or nothing that cannot be explained under the selection theory, and, at present, this theory appears to be unique in this respect."[3]

The objections of Olson and other paleontologists to the standard evolutionary theory results mainly from the fact that the fossil evidence often is missing with which to test the postulated phylogenies. Often, too, the available fossil evidence does not seem to fit comfortably into the synthetic framework, but the latter is sufficiently elastic to permit the formulation of some kind of explanation, regardless. Thus, the concept cannot be refuted, because of its high degree of flexibility and because of the impossibility of reproducing the postulated evolutionary history in the laboratory to test it. As Olson says:

> "Somehow a theory in which each case seems to be a special case fails to convey a sense of adequacy."[4]

From a different perspective, that of the systematic biologist, Ehrlich and Holm conclude:

> "Finally, consider the third question posed earlier: 'What accounts for the observed patterns in nature?' It has become fashionable to regard modern evolutionary theory as the *only* possible explanation of these patterns rather than just the best explanation that has been developed so far. It is conceivable, even likely, that what one might facetiously call a non-Euclidean theory of evolution lies over the horizon. Per-

---

[3] *Op. cit.*, p. 530.
[4] *Ibid.*, p. 540.

petuation of today's theory as dogma will not encourage progress toward more satisfactory explanations of observed phenomena."[5]

It is interesting that, in a criticism of the above position, an orthodox neo-Darwinian says:

"The proponents of strictly empirical approaches to biology have sometimes played a very useful role in restraining excesses of phylogenetic speculation by those in the Darwinian camp. But the extreme operationalist point of view not only carries with it the dangers of shallowness and superficiality but is fundamentally anti-evolutionist in emphasis, despite assertions to the contrary."[6]

The epithet of "anti-evolutionist," of course, is considered to be the acme of derision and usually has the effect of cowing any who might be bold enough even to raise a question about pronouncements of the accepted authorities. The courageous probings of men like Olson, Ehrlich, and Holm are, therefore, quite significant of what may well be an underlying discontent with much of modern evolutionary theory. An even more outspoken attack has been made by W. R. Thompson, for many years Director of the Commonwealth Institute of Biological Control at Ottawa, Canada, and a world-renowned entomologist. In defending the traditional approach to systematic classification of organisms, he says:

"Though the remark may be regarded as provocative, I must say how far superior the genetic, like the physico-chemical approach to the problems of organic form and therefore of systematics, is, to the conventional speculative approach in terms of evolutionary theory. Genetic analysis, like physico-chemical analysis, belongs to the field of positive science. Evolutionary speculation as it is commonly developed in relation to morphological and systematic problems is only too often at best merely a dressing up of comparative anatomy in such a way as to foster the illusion that we know things we do not know and are never likely to know."[7]

In a still more recent paper, Dr. Thompson severely criticizes the views of the "New Systematists," and especially of one of

---

[5] *Op. cit.*, p. 656.

[6] Grady L. Webster: "Population Biology," *Science*, Vol. 139, January 18, 1963, p. 236.

[7] "Systematics: the Ideal and the Reality," *Studia Entomologica*, Vol. 3, December, 1960, p. 498.

their most prominent spokesmen, George Gaylord Simpson, whom we have encountered before in these pages. After detailed and incisive criticisms of Simpson's efforts to reconstruct taxonomy along evolutionary lines, he says:

> "When Professor Simpson says that homology is determined by ancestry and concludes that homology is evidence of ancestry, he is using the circular argument so characteristic of evolutionary reasoning. When he adds that evolutionary developments can be described without paleontological evidence, he is attempting to revive the facile and irresponsible speculation which through so many years, under the influence of the Darwinian mythology, has impeded the advance of biology."[8]

Dr. Thompson was selected to write the foreword to the new edition of Darwin's *Origin of Species* published in the Darwinian Centennial Year as a part of the Everyman's Library Series. His entire foreword was a devastating indictment and refutation of Darwinian evolution, and perhaps even more so of the scientific morality of evolutionists! The following statement is significant:

> "As we know, there is a great divergence of opinion among biologists, not only about the causes of evolution but even about the actual process. This divergence exists because the evidence is unsatisfactory and does not permit any certain conclusion. It is therefore right and proper to draw the attention of the non-scientific public to the disagreements about evolution. But some recent remarks of evolutionists show that they think this unreasonable. This situation, where men rally to the defense of a doctrine they are unable to define scientifically, much less demonstrate with scientific rigor, attempting to maintain its credit with the public by the suppression of criticism and the elimination of difficulties, is abnormal and undesirable in science."[9]

If the reader desires a brief, yet complete refutation of all the so-called "evidences" of the theory of evolution, written by a man who is a recognized and respected biological authority, he would do well to get and read this essay by Thompson.[10]

---

[8] "Evolution and Taxonomy," *Studia Entomologica,* Vol 5, October, 1962, p. 567.

[9] Introduction to *The Origin of Species* by Charles Darwin, (New York, Everyman's Library, E. P. Dutton & Co., Inc., 1956).

[10] Thompson's Introduction has also been reprinted in the *Journal of the American Scientific Affiliation,* Vol. 12, March, 1960, pp. 2-9.

A recent book by G. A. Kerkut[11] is to the same effect, not completely rejecting evolution, but demolishing its arguments and insisting that it is not a "proved fact," as its proponents so loudly and frequently protest. Kerkut also is a recognized scientist, and whenever a recognized scientist questions evolution, cries of anger and denunciation quickly rise from the evolutionists' camp. In a review of his book, John T. Bonner says:

> "This is a book with a disturbing message; it points to some unseemly cracks in the foundations. One is disturbed because what is said gives us the uneasy feeling that we knew it for a long time deep down but were never willing to admit this even to ourselves. It is another one of those cold and uncompromising situations where the naked truth and human nature travel in different directions. The particular truth is simply that we have no reliable evidence as to the evolutionary sequence of invertebrate phyla. We do not know what group arose from what other group or whether, for instance, the transition from Protozoa occurred once, or twice, or many times.... We have all been telling our students for years not to accept any statement on its face value but to examine the evidence, and, therefore, it is rather a shock to discover that we have failed to follow our own sound advice."[12]

In spite of these admissions, however, Dr. Bonner petulantly insists that Kerkut has in no way undermined the *fact* of evolution. Theodosius Dobzhansky, another protagonist of the Neo-Darwinian school, similarly insists:

> "The basic conclusion of the author is, however, something else — since we cannot yet reconstruct in all details (sic) the phylogeny of the animal kingdom, therefore evolution is not 'proven'! This is a confusion of two distinct problems: we may be sure that life (or, for that matter, the Cosmos) had a history, but it does not follow that we know all the events of which these histories are composed with their respective dates. The author has been wise not to suggest any alternatives to the theory of evolution: ..."[13]

It is not too well known in this country that Darwinian evo-

---

[11] *Implications of Evolution* (New York, Pergamon Press, 1960) 174 pp.

[12] Review of Kerkut's book, *American Scientist,* Vol. 49, June, 1961, p. 240.

[13] Review of Kerkut's book, *Science,* Vol. 133, March 17, 1961, p. 753.

lution has been under intense attack by scientists in France for quite a few years. A recent review of this French discussion says:

> "This year saw the controversy rapidly growing, until recently it culminated in the title 'Should We Burn Darwin?' spread over two pages of the magazine *Science et Vie*.
>
> "The article, by the science writer Aime Michel, was based on the author's interviews with such specialists as Mrs. Andree Tetry, professor at the famous *Ecole des Hautes Etudes* and a world authority on problems of evolution, Professor Rene Chauvin and other noted French biologists, and on his thorough study of some 600 pages of biological data collected, in collaboration with Mrs. Tetry, by the late Michael Cuenot, a biologist of international fame.
>
> "Aime Michel's conclusion is significant: the classical theory of evolution in its strict sense belongs to the past. Even if they do not publicly take a definite stand, almost all French specialists hold today strong mental reservations as to the validity of natural selection."[14]

This does not mean, of course, that these French biologists have rejected the *"fact"* of evolution. But evidently the best mechanism they can suggest is that "genetic change may be... controlled by the *good judgment* of the organism itself"(!).[15]

Thus, there is good reason to believe that more and more qualified scholars today are recognizing that the scientific basis of the evolution theory is very weak, and many are rejecting it entirely. Personally, after talking with hundreds of evolutionists and after reading hundreds, probably thousands, of evolutionary books and articles, the writer is more and more convinced of two facts: (1) there is not one shred of genuine evidence, either in science or Scripture, for the validity of evolution; (2) the only reason why most people seem to believe in evolution is either because they *want* to believe in it or else because they have been cowed into accepting it out of fear of being called ignorant or reactionary or some such fearful name.

A great many people, particularly intellectuals, simply *prefer* an evolutionary theory of origins, because this device either

---

[14] Zygmunt Litynski: "Should We Burn Darwin?", *Science Digest*, Vol. 51. January 1961, p. 61.

[15] *Ibid*, p. 63.

consciously or subconsciously relegates the Creator to a far-off, indefinite, or even illusory, role in the universe and in the lives of men who are in moral rebellion against him. They have simply rejected God and their responsibility to him, and so scoff: "Where is the promise of his coming?" They prefer to "worship and serve the creature rather than the Creator."

Many others, including many who sincerely desire to know God and his truth, believe in evolution for the simple reason that they think science has proved it to be a "fact" and, therefore, it must be accepted. They know relatively little about either the scientific or philosophical questions involved, never having bothered to make a study of them, and have been persuaded by those in the first group that "all educated people must accept evolution." In recent years, a great many people in this group, having finally been persuaded to make a real examination of the problem of evolution, have become convinced of its fallacy and are now convinced anti-evolutionists. The writer would himself be included in this group.

But behind both groups of evolutionists one can discern the malignant influence of "that old serpent, called the Devil, and Satan, which deceiveth the whole world" (Revelation 12:9). As we have seen, it must have been essentially the deception of evolution which prompted Satan himself to rebel against God, and it was essentially the same great lie with which he deceived Eve, and with which he has continued to "deceive the whole world."

Consequently, until Satan himself is destroyed, we have no hope of defeating the theory of evolution, except intermittently and locally and temporarily. It is not the fact that many here and there are awakening to the falseness of evolution, or to the weaknesses in its scientific base, that encourages us to think we are in evolution's twilight period. Neither is it the fact that more and more people are becoming aware of the real issues and are having to make a choice between creation and evolution, with all the implications stemming from this choice.

But we can speak confidently of the imminent death of evolution because we can discern ample signs of the imminent "coming" of the Lord, which the latter-day scoffers so vigorously resist. The very fact that uniformitarian and evolutionary thought seems to have captured the intellectual world is noted by the Apostle Peter (II Peter 3:3, 4) as indicative of the "last days." Similarly, the Apostle Paul, in the last epistle written before his martyrdom, emphasized that "in the last times," men would be "having a form of godliness, but denying the power thereof" (II Timothy 3:1, 5). That is, they would profess some kind of religion or morality, but would deny that there was any "power" to it, and this is nothing else than the denial of the miraculous, of creation, of His coming in judgment, and so again is essentially an allegiance to the concept of uniformity and, therefore, evolution.

The culmination of all rebellion against the Creator, both on the part of man and of Satan, will ultimately be a unified and world-wide system of humanistic culture, religion and government. As noted in the first chapter, practically all disciplines of modern thought are now structured around the evolutionary philosophy. That is to say, essentially all of modern society and its culture is man-centered, rather than God-centered. Man has "created" God in his own image, so that man considers himself capable of working out his own salvation and that of human society. Man is the pinnacle of all the evolutionary processes of the ages, and now is capable of bringing about further evolution as needed for the perfectibility of himself and his environment.

This monstrous system, toward which numerous world movements seem now rapidly to be gravitating (e.g., international communism, the ecumenical movement in religion, world socialism, the United Nations and its multitudinous tentacles, etc.) seems all but certain to culminate sooner or later in the Biblical Antichrist, which will be both a world-system and the Satan-inspired man at the pinnacle of that system. Of this man, Scripture says:

> "And he opened his mouth in blasphemy against God, to blaspheme his name, and his tabernacle, and them that dwell in heaven. And it was given unto him to make war with the saints, and to overcome them: and power was given unto him over all kindreds, and tongues, and nations. And all that dwell upon the earth shall worship him, whose names are not

> written in the book of life of the Lamb slain from the foundation of the world. If any man have an ear, let him hear" (Revelation 13:6-9).

If anyone should wonder how it is, in this age of science and enlightenment, that all the world should actually *worship* a man, or a man-system, let him reflect how it is that in this 20th century millions have all but worshipped Lenin, Stalin, and Mao; and how other millions have all but worshipped Mussolini, and Hitler, and Hirohito, and Nkrumah, and Gandhi, and Father Divine, and assorted other latter-day "deities"; and then the possibility will not sound so far-fetched after all. Add to this the well-nigh universal teaching of evolution with its implication that the cosmic process has "come to consciousness" in man, the liberal religious dogma of the divinity of man, the psychological rejection of the reality of sin, and all the other aspects of modern man-centered culture; and the imminence of an all-powerful state, with some great representative of mankind at its head, worshipped by all men (who thereby in reality are worshipping themselves in him), assumes an aspect of high probability.

We do not propose to speculate on how this will be brought about, although there are certain intimations in Scripture, but it is evident that there are numerous very powerful movements in the world today which are directed to just such a goal. And that it *will* come is plainly stated in the Bible:

> "Let no man deceive you by any means: for that day shall not come [that is, the day of the Lord], except there come a falling away [literally, *the apostasy]* first, and that man of sin be revealed, the son of perdition; who opposeth and exalteth himself above all that is called God, or that is worshipped; so that he as God sitteth in the temple of God, showing himself that he is God" (II Thessalonians 2:3, 4).

This world-system and its Premier, when they come, will be especially prepared and energized by Satan himself, with the purpose of one last all-out assault against the throne of God. "Woe to the inhabiters of the earth and of the sea! for the devil is come down unto you, having great wrath, because he knoweth that he hath but a short time" (Revelation 12:12).

Finally, this world religio-political power will become openly and blatantly the final assault of Satan against God. It will no longer be necessary to camouflage his rebellion as "humanism," or "socialism," or some other movement designed to glorify man. Men will have finally and knowingly and irrevocably rejected God and his Christ.

> "The kings of the earth set themselves, and the rulers take counsel together, against the Lord, and against his Christ, saying, Let us break their bands asunder, and cast away their cords from us" (Psalm 2:2, 3).

Men will then come under the full spell of the great delusion which has energized Satan through all the ages, thinking that God is of the same order as his creatures and can ultimately be defeated under the leadership of Satan.

> "And then shall that Wicked [One] be revealed.... whose coming is after the working of Satan with all power and signs and lying wonders, and with all deceivableness of unrighteousness in them that perish; because they received not the love of the truth, that they might be saved. And for this cause, God shall send them strong delusion, that they should believe a lie" (II Thessalonians 2:8-11).

Men will come to acknowledge the great Man, the Antichrist (also called the Beast in the Bible), as God. But through him, they will sooner or later see that in reality they are worshipping Satan who has supernaturally endowed and equipped this personage with his tremendous abilities and power, and they will enthusiastically follow the devil in his all-out drive to dethrone God.

> "and the dragon gave him his power and his seat, and great authority.... and all the world wondered after the beast. And *they worshipped the dragon* which gave power unto the beast: and they worshipped the beast, saying Who is like unto the beast? who is able to make war with him?" (Revelation 13:2-4).

Perhaps the Russian communist Zinonieff was not speaking metaphorically at all when he boasted: *"We shall grapple with the Lord God. In due time we shall vanquish him from the highest Heaven, and where he seeks refuge, we shall subdue him forever."*

But in spite of the proud boastings of Satan and all the multitudes of men who have followed him in his great delusion, God and his Christ will triumph one day, forever! Then will be the death of evolution and unbelief and pride, and of all the systems of man which have been built upon these things. Speaking of the final confederation of nations of the world and their united opposition to God, the Bible says:

> "These have one mind, and shall give their power and strength unto the beast. These shall make war with the Lamb, and the Lamb shall overcome them: for he is Lord of lords, and King of kings: and they that are with him are called, and chosen, and faithful" (Revelation 17:13, 14).

It is intensely significant that, in the context of the terrible prophecies concerning this final rebellion and its consummation, it is recorded that God sends forth an angel to preach the *everlasting gospel* to all the kindreds and nations and tongues and peoples of the earth, with one last call to repentance and acknowledgment of God as true God and Creator. It is of marvelous timeliness that the burden of this everlasting gospel (and of course there is only *one* gospel, the good news of atonement and redemption and reconciliation of man to God through the death and resurrection of the Lord Jesus Christ), is centered around the theme of the *fact* of creation, implying that no one really accepts Jesus Christ and his salvation who does not also accept him as Creator. The angel says:

> "Fear God, and give glory to him; for the hour of his judgment is come: and worship him that made heaven and earth, and the sea, and the fountains of waters" (Revelation 14:7).

Index of Subjects

Index of Names

Index of Scriptures